PLANNING AND ANALYSIS OF CONSTRUCTION OPERATIONS

PLANNING AND ANALYSIS OF CONSTRUCTION OPERATIONS

by

Daniel W. Halpin, Ph.D.
Purdue University

and

Leland S. Riggs, Ph.D.
Georgia Institute of Technology

A WILEY INTERSCIENCE PUBLICATION

JOHN WILEY & SONS, INC.

New York   •   Chichester   •   Brisbane   •   Toronto   •   Singapore

In recognition of the importance of preserving what has been
written, it is a policy of John Wiley & Sons, Inc., to have books
of enduring value published in the United States printed on
acid-free paper, and we exert our best efforts to that end.

Library of Congress Cataloging-in-Publication Data

Halpin, Daniel W.
     Planning and analysis of construction operations / by Daniel W.
Halpin and Leland S. Riggs.
        p.   cm.
     "Based upon and is an extension of the original text entitled
"Design of construction and process operations" by Daniel W. Halpin
and Ronald W. Woodhead published in 1976"—Pref.
     Includes bibliographical references (p.   ) and index.
     ISBN 0-471-55510-X : $54.95
     1. Contractors' operations—Computer simulation.   2. Contractors'
operations—Management.   I. Riggs, Leland S.   II. Halpin, Daniel W.
Design of construction and process operations.   III. Title.
TA210.H25   1992
624'.068—dc20   91-32231
                                                           CIP

PREFACE

This book is based on and is an extension of the original text entitled *Design of Construction and Process Operations* by Daniel W. Halpin and Ronald W. Woodhead published in 1976. The authors would like to acknowledge that a number of chapters and figures are taken directly from that text and thank Ronald W. Woodhead for his generosity in allowing use of this material from the original book.

Since the publication of the original text, the concept of simulation as a method of analyzing construction operations has been added to the curricula of at least 50 university-level construction programs around the world. An extensive literature relating to the modeling of construction operations using the CYCLONE (CYCLic Operations NETwork) approach has developed. Moreover, the interest in automating construction operations at the production level during the past decade has led to a search for better and more effective ways of modeling construction processes and technologies as they are used in the field.

New technologies in the construction field have become more and more based on the concepts of repetitive construction sequences. Modularization and the use of repetitive features in all kinds of construction have led to more efficient use of materials, manpower, and machines.

The objective of this text is to introduce concepts that form the basis for a scientific or analytical approach to the subject area of design and evaluation of productive processes as they are encountered in construction. The time is ripe for a science of construction operations. This text builds on techniques that were originally developed in the previous book. The CYCLONE modeling system provides a quantitative way of viewing, planning, analyzing, and controlling construction processes and operations. This system provides a framework in which operations can be reduced to simple flowcharts that help in studying the interaction of the resources required as well as the determination of rates of production at the work site.

Supporting techniques such as queueing theory and line-of-balance methods are presented to provide a background to the student of quantitative modeling methods. The main body of the book focuses on the use of simulation techniques to model and evaluate repetitive construction operations. A wide variety of examples from all areas of construction are presented to demonstrate the power and simplicity of simulation in analyzing production of complex construction operations. Methods of balancing production rates among interfacing processes and optimally allocating resources for maximum production are presented.

The book is divided into several sections. The first four chapters introduce techniques that are helpful in learning to break construction processes into work tasks.

The next four chapters introduce the basic modeling concepts of the CYCLONE system. Chapters 9 and 10 discuss the basic concepts of next-event discrete simulation and how it can be used to evaluate the production of processes and the utilization of the interacting resources. Chapters 11–13 discuss the modeling of a wide variety of construction operations using the CYCLONE modeling framework. The final two chapters of the text address the details of applying simulation techniques to evaluation of actual construction operations.

A microcomputer program called MicroCYCLONE is available to support the use of the CYCLONE system in the classroom environment. This program is described in one of the appendices. Appendixes A and B present material in support of the queueing theory concepts presented in Chapter 3.

The authors would like to acknowledge the contributions of numerous students who have worked with the techniques presented over the past 18 years. Major contributions to the final text of this book have been made by Simaan AbouRizk, James Lutz, and Dennis McCahill. Many of the models discussed in the text were developed originally as classroom projects by students at Georgia Institute of Technology, the University of Maryland, and Purdue University. The authors would also like to acknowledge the support of Rachel Haas, Antonio Gonzalez, Roberto Eljaiek, and Ted Vrehas in preparing the final manuscript.

DANIEL W. HALPIN

West Lafayette, Indiana
January 1992

LELAND S. RIGGS

Atlanta, Georgia

CONTENTS

1 Construction Processes

1.1 THE NATURE OF CONSTRUCTION OPERATIONS

Construction requires the application of a diverse palette of resources to realize a finished facility—a high-rise building, an airport runway, an alpine bridge, or a subsurface tunnel in an urban area. The organization and application of these resources can be viewed in terms of the level at which decisions are being made; that is, there is a construction hierarchy that is dictated by the way in which construction is organized. At the company level, decisions related to which projects to bid and the recruitment of personnel are of interest. At the project level, decisions regarding how long it will take to complete a facility and the selection and movement of resources such as machines and workers must be considered. Ultimately, however, the project must be constructed. Physical items such as concrete, gypsum board, glass, steel, and a broad spectrum of materials must be erected, placed, and installed to achieve the completed facility. This is the production level in construction. This is where planning and design, analysis, and control measures come together to realize the end item—the facility.

One of the most critical questions confronted by a construction engineer is "What construction technique should be selected?" The number and types of methods available for construction are diverse. New technologies are continuously being adapted for use in the construction process and lead to an expanding variety of construction techniques. Many of the construction methods presently being used did not exist 30 years ago. The postwar era has seen the development of a staggering variety of products and processes that are changing the way in which buildings, roads, bridges, and tunnels are being constructed. The development of new methods and the adaptation of new and advanced technology to the construction arena has led to a greater concentration on improved methods for analysis and design of construction processes.

1.2 THE HIERARCHY OF CONSTRUCTION

As noted above, organizational considerations lead to a number of hierarchical levels that can be identified in construction. This derives from the fact that construction is traditionally carried out in a project format. Decisionmaking at levels above the project relate to company management considerations. Decisions within the project relate to operational considerations (e.g., selection of production meth-

1

ods) as well as the application of resources to the various construction production processes and work tasks selected to realize the constructed facility.

Specifically, four levels of hierarchy can be identified as follows:

1. *Organizational.* The organizational level is concerned with the legal and business structure of a firm, the various functional areas of management, and the interaction between head office and field agents performing these management functions. The organizational level is not considered in this book.

2. *Project.* Project level vocabulary is dominated by terms relating to the breakdown of the project for the purpose of time and cost control (e.g., *the project activity and the project cost account*). Also, the concept of resources is defined and related to the activity as either an added descriptive attribute of the activity or for resource scheduling purposes.

3. *Operation (and Process).* The construction operation and process level is concerned with the technology and details of how construction is performed. It focuses on work at the field level. Usually a construction operation is so complex that it encompasses several distinct processes, each having its own technology and work task sequences. However, for simple situations involving a single process, the terms are synonymous.

4. *Task.* The task level is concerned with the identification and assignment of elemental portions of work to field agents.

The relative hierarchical breakout and description of these levels in construction management are shown in Figure 1.1. It is clear that the organizational, project, and activity levels have a basic project and top management focus, while the operation, process, and work task levels have a basic work focus. In order to develop definitions that distinguish between terms used at these levels, it is important to start at the work task level.

A *work task* is the basic descriptive unit in construction practice. If a work task is broken down into components, human factor considerations or detailed equipment motions are involved that assume microanalysis of motions and actions.

A work task should be a readily identifiable component of a construction process or operation. Its description must be so clear that any member of a construction crew can readily grasp and visualize what is involved in and required of the work task. *Work tasks are therefore the basic building blocks of processes and operations.*

A *work assignment* is the collection of work tasks specifically assigned to a crew member for performance. Work assignments usually involve sequences of work tasks appropriate to a certain trade and skill level of the worker and, therefore, they may define a construction process.

A *construction process is defined as a unique collection of work tasks related to each other through a technologic structure and sequence.* A construction process therefore represents a technologic or readily identifiable segment of a construction operation. It may represent the impact on an operation of a member or trade component of a mixed crew. It may, however, represent one of many pro-

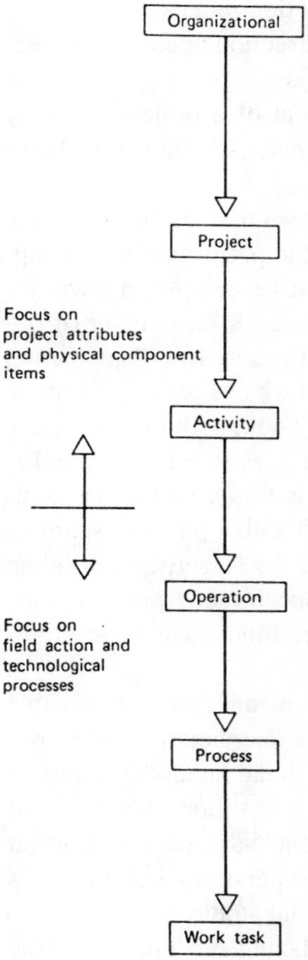

Hierarchical Level   Description and Basic Focus

Organizational	Company structure and business focus. Head office and field functions. Portfolio of projects. Gross project attributes: total cost, duration, profit, cash flow, percent complete.
Project	Project definition, contract, drawings, specifications. Product definition and breakdown into project activities. Cost, time, and resource control focus.
Activity	Attainment of physical segment of project equated to time and cost control. Current cost, time, resource use. Status focus.
Operation	Construction method focus. Means of achieving construction. Complete itemized resource list. Synthesis of work processes.
Process	Basic technological sequence focus. Logical collection of work tasks. Individual and mixed trade actions. Recognizable portion of construction operation.
Work task	Fundamental field action and work unit focus. Intrinsic knowledge and skill at crew member level. Basis of work assignment to labor.

Focus on project attributes and physical component items

Focus on field action and technological processes

Figure 1.1 Hierarchical levels in construction management.

cesses that a single worker can perform because of personal trade training and skill. In this sense a work assignment to a crew member may involve that person in different construction processes at different times or at different sequences in a construction operation. Many construction processes are highly repetitive: an operator continuously cycles through work tasks processing or using resources, thereby achieving progress in construction.

A construction operation results in the placement of a definable piece of construction and inherently has some technologic processes and work assignment structure.

The construction operation is closely related to the means of achieving an end product and can be repetitive in nature. It is therefore an expression of and linked to a construction method and only indirectly with a physical segment of the completed product. It is, in fact, a synthesis of construction processes. In simple operations the operation is identical with its single construction process.

The timeframe duration of the basic cycle of the construction operation is measured in hours or days.

An *activity* is a time- and resource-consuming element of a project normally defined for the purpose of time and cost control by a planner, estimator, scheduler, or cost engineer.

Activities are the aggregation of operations or processes that contribute to the completion of a physical component of the structure or the performance of a support service. In this sense an activity is unique and must be completed once, although its completion may require the repeated performance of a number of operations or processes, some of which may be unique to the activity. Construction operations or processes are often common to many activities and assume a unique magnitude and significance for a specific activity. In this book, activities represent a significant portion of the project and have a duration of days, weeks, or months.

The distinction between an activity and an operation is firmly tied to the duration of the function and whether it is primarily concerned with a physical segment unique to the time cost control of the project (i.e., as it is for an activity) or to the technological process or method required to achieve a specific end product (i.e., as it is for an operation). The operation is therefore more fundamental for an understanding of field methods.

As mentioned above, further breakouts of construction functions are possible at lower microlevels or higher macrolevels. Sometimes a microanalysis of a construction operation or process is useful and is undertaken in the field when a highly repetitive process is required for a work item of large magnitude. Although all construction projects are unique and few are of sufficient magnitude to warrant special studies, contractors often fail to realize that many operations and processes are common to every project and, therefore, require special study.

To illustrate the definitions given above, consider a glazing subcontract for the installation of glass and exterior opaque panels on the four concourses of the Hartsfield International Airport in Atlanta, Georgia. This was a highly repetitive process requiring the installation of five panels per bay on 72 bays of each of the four concourses. Figure 1.2 shows a schematic diagram of the project. A breakout of typical items of activity at each level of hierarchy is given in Table 1.1. At the project level, activities within the schedule relate to the glass and panel installation in certain areas of the concourses. At the work task level, location, unloading, stripping, and other crew-related activity is required.

1.3 REPETITIVE PROCESSES IN CONSTRUCTION

Although constructed facilities themselves are typically unique, the methods used to construct them often are repetitive or cyclic in nature as in the case of the

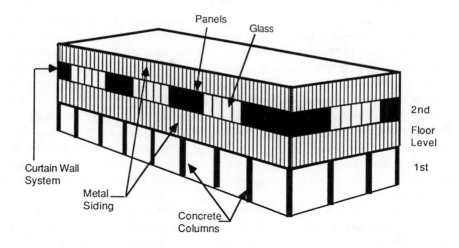

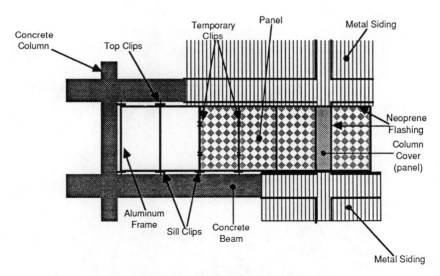

Figure 1.2 Schematic of concourse building.

TABLE 1.1   Glazing Example of Hierarchical Terms

Project	Installation of all exterior glass and panel wall construction on Concourses A–D of the Hartsfield International Airport, Atlanta, GA
Activity	Glass and panel installation on Concourse A, Bays 65–72
Operation	Frame installation to include preparation and installation of 5 panel frames in each concourse bay; Column cover plate installation
Process	Sill clip placement; mullion strips installation
	Glass placement in frame; move and adjust hanging scaffold
Work task	Locate and drill clip fastener; unload and position mullion strips; strip protective cover from glass panel; secure scaffold in travel position

Hartsfield Airport glazing project. Construction techniques can be roughly categorized as *cyclic* or *noncyclic*. Cyclic processes usually depend on the repetition of tasks required to produce a part of the finished project. A noncyclic process would be a one-time sequence that is not repeated. The lifting of a large boiler into position is a nonrepetitive process. On the other hand, the forming of floor slabs in a tall building tends to be repetitive in nature (see Fig. 1.3).

Projects that evidence a linear nature often depend on cyclic or repetitive sequences of work. Typical examples are the paving of a roadway, the driving of a tunnel, and the placement of precast panels on a high-rise building. You can probably think of other construction processes that require the repetition of a sequence of tasks. Clearly, repetition is desirable since it leads to efficiency of resource utilization.

In manufacturing, the cornerstone of mass production is repetitiveness of the work to be performed. This is based on the standardization of the product to be created. Standardization and modularization are historically well-known concepts

Figure 1.3 Repetitive forming in high-rise construction.

for construction materials (e.g., brick and block sizes, dry wall panels). The concept of standardization to achieve repetition has been less successfully applied in the design of construction processes.

Recently, successes on large projects have proved that design of processes to achieve repetitiveness is the basis for cost-effective construction, which also leads to high quality. The advantage of economies-of-scale that result when the work is modularized and repetition is increased have been well documented (e.g., segmented bridging, off-site prefabrication).

1.4 REPETITION BASED ON ADVANCED CONSTRUCTION TECHNOLOGY

The Val Crotta Bridge constructed in the south of Switzerland is a good example of how new design concepts utilizing advanced technology lead to repetition in construction. The bridge is 145 m long, with a center arch of 92 m. The approach spans are 9.5 and 14.5 m on one side and 15.5 and 13.5 m on the other. The deck section carrying the roadway is a slab construction supported by two large girders.

The substructure consists of two parallel arches each with a 1-m width. The depth of each arch varies from 1.3 to 1.8 m. The arch support is 2.1 m lower on one side. The two arches are 2.2 m apart, and cross-bracing at the crown is 80 cm deep while the cross connectors at the arch supports are 160 cm deep.

The roadway is approximately 70 m above the lowest point of the valley. Since the use of falsework to support the casting of the arches would have been prohibitively expensive, an alternative support system was specified by the designer. This cable-supported system led to a repetitive jump forming method for constructing the arches. This system is shown schematically in Figure 1.4.

1.5 CABLE-SUPPORTED FORMING SYSTEMS

When inclined cable support systems are considered, it is normally in connection with the support of permanent bridge structures with spans in excess of 200 m. The suspension bridge across the Rhine River in Cologne is a well-known example of an inclined cable-support structure. The Sunshine Skyway Bridge connecting St. Petersburg and Bradenton, Florida, is also an example of this method of support. This notwithstanding, a large number of smaller structures have been built using this technique as a permanent or temporary support system.

Figure 1.5 shows how the inclined cable approach was used to cast the bridge arch in place. First pylons were constructed atop the approach ramps on each side of the valley high enough to support cables supporting the forming and (subsequently) the cast-in-place arch segments. As can be seen from the figure, seven cables were supported from each pylon. Eight segments are cast sequentially on each side. Starting at each arch support, the jump form is advanced to a new

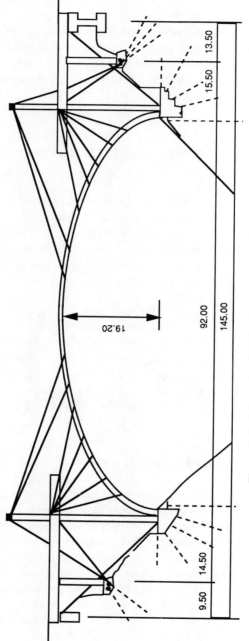

Figure 1.4 Schematic of inclined cable construction procedure.

Figure 1.5   Work on cable-supported casting system.

location and prepared for casting. The previous steps are repeated on each side until all segments have been cast and supported except the crown (key) segment. This segment is cast last, and the structural integrity of the arch is realized. Once the crown segment has achieved sufficient strength, the inclined cable supports can be removed. Portions of this process are shown in Figure 1.4.

The bid price submitted for this work was substantially less than the price required had conventional falsework support been used. This incremental approach reduced significantly the number of labor hours required and the size of the crew required. The precision of the casting was ensured by careful survey control throughout the casting process. As each segment was formed and cast, measurements were made to ensure that the two half arches would mate properly at the center.

Using an extension of this approach, an Austrian firm constructed an arch bridge over the Argens River in southern Bavaria. Two gigantic half arches were "prefabricated" in a vertical position on the arch supports using jump forming procedures to form them as curved columns. The position of the half arches was maintained as the forming sequence advanced to keep the form moving vertically. The precise curvature of the half arches was maintained by lasers at the arch supports used to monitor the operation. Once casting was complete, the half-arches were then lowered into place using a computer-controlled cable-support system until the full arch was formed. The final 32.2-m-deep arch supports a bridge 232.3 m in length (see "Arch Halves Fall . . . ," *Engineering News-Record*, June 21, 1985.)

1.6   MODELING CONSTRUCTION PROCESSES

A model is a representation of a real-world situation and usually provides a framework with which a given situation can be investigated and analyzed. Models contain or portray data about a situation that, when interpreted according to certain rules or conventions, provide information about the situation relevant to pertinent decision processes. Maps constitute a modeling of roads, rivers, and physical terrain (e.g., a geographic modeling). The globe of the earth is a model that we commonly refer to in order to organize our ideas regarding the location of geographic features.

In the study of hydrologic basins, models are frequently used. Portions of the Mississippi River are modeled to a high level of detail at the Waterways Experiment Station in Vicksburg, Mississippi. This allows hydrologists to study the impact of structures constructed along the river on erosion, silting, and the change of the channel. This is a higher-level application of mapping to perform analytic studies.

In certain cases, full-scale models are used for training purposes. During World War II, trainers replicating the interior of an aircraft cockpit complete with controls and instruments were developed for pilot training. These Link trainers were invaluable in training pilots as to how to react to various emergency situations as well as how to fly on instrumentation during bad weather. This concept is still used for commercial pilot certification using replicas of the cockpits of various commercial aircraft. An extension of this full-scale modeling is used for NASA (National Aeronautics and Space Administration) astronaut training.

The level of abstraction of the model from the situation being modeled varies and most analytic models are abstractions of the real world. The globe, for instance, is really an abstraction of the real world (although we accept it as being an accurate representation). Many physical systems can be modeled by mathematical abstractions of the actual system. The analysis of the structural frame of a high-rise building is a mathematical abstraction. We assume the building to be an interconnected network of linear elements (e.g., beams and columns). This abstraction allows us to calculate an acceptable approximation of the shears and moments developed throughout the frame when it is subjected to a variety of loading conditions.

Mechanical engineers utilize modeling analogs to investigate mechanical systems. The shock absorbers in a vehicle and their response to various input phenomena can be modeled as electrical circuits of resistance, inductance, and varying current flows. These abstractions can be thought of as conceptual models that are handy for solving real-world problems.

In construction management, mathematical models are helpful in addressing problems of planning and control. Networks subject to simple mathematical solutions are commonly used for project scheduling. Cash flow throughout the life of a project can be predicted using mathematical models. Three-dimensional computer models of large industrial plants are being widely used for design and construction purposes.

The precision with which these models reflect the real world varies widely. Abstract or conceptual models depend on a set of modeling and interpretive rules. Scheduling networks and bar charts, for example, have their own individual modeling and interpretive rules. Schematic models are representations that portray a physical situation (e.g., a map) so that a physical modeling reaction or perception is induced in the user. Construction drawings are an excellent example of a schematic model. Models of various types are at the heart of problem-solving. Reasoning in general depends on establishing a modeling rationale.

This book deals with various modeling techniques that are of interest in solving problems at the process (i.e., the production) level in construction. These are quantitative models in the sense that they require numeric input to represent the situation being modeled and then provide numeric output that must be analyzed to understand the response of the system being modeled. The following sections will discuss some of the quantitative modeling techniques available to the construction manager.

1.7 DETERMINISTIC MODELS

Simple deterministic models have been developed for earthmoving operations based on rated equipment characteristics, equivalent grades, and haul distances. These models are described as deterministic since the time durations involved are assumed to be fixed or constant values that do not vary over time. Any variability in the process work task durations is assumed to be small or insignificant.

To illustrate this, consider a simple earthmoving haul operation. A diagram of the cycles involved is shown in Figure 1.6. The haul unit is a 30-yd^3 scraper and it is loaded in the cut area with the aid of a pusher dozer. A simple calculation of the production of this two-cycle system can be made using deterministic work task durations.

To determine the deterministic durations for the scraper travel times to and from the fill location, it is necessary to consult the performance handbooks published by most manufacturers; these handbooks show the development of this information from charts or specifically developed nomographs. Assume that in this case the 30-yd^3 tractor scraper is carrying rated capacity and operating on a 3000-ft-level

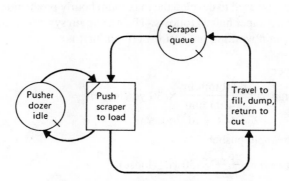

Figure 1.6 Scraper–pusher dual-cycle model.

haul where the rolling resistance (*RR*) developed by the road surface is 40 pounds per ton. Using a standard formula, this converts to

$$\text{Effective grade} = \frac{RR}{20 \text{ lb/ton/\% grade}} = \frac{40 \text{ lb/ton}}{20 \text{ lb/ton/\% grade}} = 2\%$$

By consulting the nomographs given in Figure 1.7, the following travel times can be established:

$$\text{Time loaded to fill} \quad = 1.4 \text{ min}$$

$$\text{Time empty to return} = 1.2 \text{ min}$$

Assume further that the dump time for the scraper is 0.5 min and the push time using a 385-horsepower (hp) track-type pusher tractor is 1.23 min, developed as follows (see, for instance, Caterpillar Tractor Company publications).

$$\text{Load time} \quad = 0.70$$

$$\text{Boost time} \quad = 0.15$$

$$\text{Transfer time} = 0.10$$

$$\text{Return time} \quad = \underline{0.28}$$

$$\text{Total} \quad\quad = 1.23 \text{ min}$$

Using these deterministic times for the two types of flow units in this system (i.e., the pusher and the scrapers) the scraper and pusher cycle times can be developed, as shown in Figure 1.8 as follows:

$$\text{Pusher cycle} \quad = 1.23 \text{ min}$$

$$\text{Scraper cycle} = 0.95 + 1.2 + 1.4 + 0.5 = 4.05 \text{ min}$$

These figures can be used to develop the maximum hourly production for the pusher unit and for each scraper unit as follows. The maximum system productivity (Prod) [assuming a 60-min working hour, (i.e., 60 min/hr)] is

1. Per scraper

$$\text{Prod (scraper)} = \frac{60 \text{ min/hr}}{4.05 \text{ min}} \times 30 \text{ yd}^3 \text{(loose)}$$
$$= 444.4 \text{ yd}^3 \text{ loose/hr}$$

2. Based on single pusher

$$\text{Prod (pusher)} = \frac{60}{1.23} \times 30 \text{ yd}^3 \text{ (loose)}$$
$$= 1463.4 \text{ yd}^3 \text{ loose/hr}$$

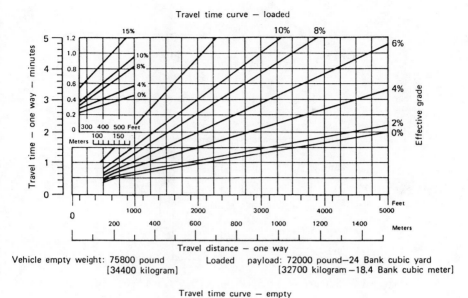

Travel time curve — loaded

Vehicle empty weight: 75800 pound   Loaded   payload: 72000 pound—24 Bank cubic yard
          [34400 kilogram]   [32700 kilogram —18.4 Bank cubic meter]

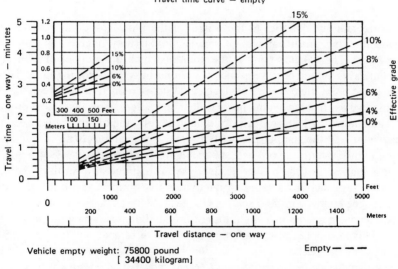

Travel time curve — empty

Vehicle empty weight: 75800 pound   Empty — — —
          [34400 kilogram]

Figure 1.7 Travel time nomographs (Caterpillar Tractor Co.).

Using these productivities based on a 60-min working hour, it can be seen that the pusher is much more productive than a single scraper and would be idle most of the time if matched to only one scraper. By using a graphical plot, one can determine the number of scrapers that are needed to keep the pusher busy at all times.

The linear plot of Figure 1.9 shows the increasing productivity of the system as the number of scrapers is increased. The productivity of the single pusher constrains the total productivity of the system to 1463.4 yd^3. This is shown by the

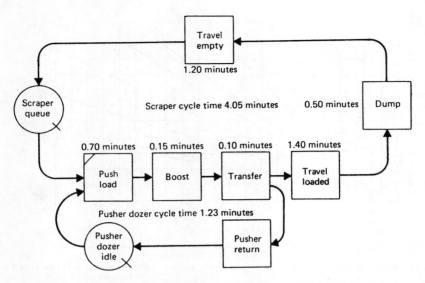

Figure 1.8 Scraper–pusher cycle times (Halpin and Woodhead 1976).

dotted horizontal line parallel to the x axis of the plot. The point at which the horizontal line and the linear plot of scraper productivity intersect is called the *balance point*. The balance point is the point at which the number of haul units (i.e., scrapers) is sufficient to keep the pusher unit busy 100% of the time. To the left of balance point, there is an imbalance in system productivity between the two interacting cycles; this leaves the pusher idle. This idleness results in lost productivity. The amount of lost productivity is indicated by the difference between the horizontal line and the scraper productivity line. For example, with two scrapers operating in the system, the ordinate AB of Figure 1.9 indicates that 574.6 yd^3, or a little less than half of the pusher productivity, is lost due as a result of the mismatch between pusher and scraper productivities. As scrapers are added, this mismatch is reduced until, with four scrapers in the system, the pusher is fully utilized. Now the mismatch results in a slight loss of productivity caused by idleness of the scrapers. This results because, in certain instances, a scraper will have to wait to be loaded until the pusher is free from loading a preceding unit. If five scraper units operate in the system, the ordinate CD indicates that the loss in the productive capacity of the scraper because of delay in being push loaded is

$$\text{Productive loss} = 5 \ (444.4) - 1463.4 = 758.6 \ \text{yd}^3.$$

This results because the greater number of scrapers causes delays in the scraper queue for longer periods of time. The imbalance or mismatch between units in dual-cycle system resulting from deterministic times associated with unit activities is called *interference*. It is due only to the time imbalance between the interacting cycles. It does not consider idleness or loss of productivity because of random variation in the system activity durations.

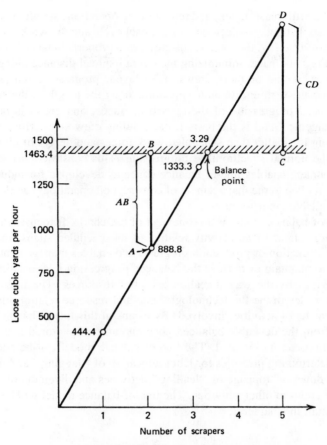

Figure 1.9 Productivity plot.

1.8 LINE-OF-BALANCE MODELS

As discussed previously, in certain types of projects repetitive sequences of activities develop from the nature of the project and the construction technology adopted. In such cases the individual activities must be properly synchronized in duration and productivity if gross delays and misuse of resources are to be avoided. Often mishaps and delays in working the activities compound the instability between the sequence of activities, even in properly balanced situations, with consequent loss of overall productivity and the generation of chain-reaction management problems (e.g., ripple effects).

Typical examples of projects involving repetitive sequences of activities are high-rise buildings, tunnels, and pipelines. In high-rise buildings repetitive activity sequences develop from the nature of floor-to-floor operations. Thus formwork erection, steel rebar placement, and concreting interact and influence the following "close-in" trades working on exterior walls and window panels. Similarly, in tunneling, the drill, blast, muck, and tunnel lining operations are often built around a shift basis so that delays affect the entire tunnel cycle and the work force. Again, in pipeline construction, the individuals crews of a properly balanced spread move forward at the same pace, minimizing the overall spread distance along the pipeline; however, if one or more crews are delayed, progressive individual crew downtimes occur, with uneconomic spreading along the length of the pipeline.

The balanced progression of the activities, trades, and crews is desirable to project management and is the basis for scheduling crew sizes. However, variations resulting in smaller than scheduled work forces and delays of all types tend to perturb the normal synchronized progression. This construction and management environment enables line-of-balance concepts developed for industrial situations to be applied to the management of construction operations, work tasks, and the daily available labor resources.

The line-of-balance model is a refinement of bar charts. It focuses on monitoring the current status of an activity relative to its scheduled status, incorporates updating as a function of production progress, and enables management decisions to be made to maintain or retrieve the balanced progression relationships between project activities by the way it reallocates labor resources. The line-of-balance approach is to determine the technologic time and resource relationship between the sequential set of activities involved. By means of this line of balance, relative departures from the desirable balanced position can be monitored and corrective remedial action can be initiated. The line-of-balance model can be considered a means for determining priorities for labor allocation of ''lagging'' activities or for the slowing down or stoppage of ''leading'' activities and diversion of their labor resources or crews to other activities. The line-of-balance model will be discussed in greater detail in Chapter 2.

1.9 QUEUEING MODELS

Many situations in which units are repetitively processed can be viewed as queueing or waiting-line problems. Queueing models can be used to model systems in which two units, designated as *processor* and *calling unit*, interact with one another. Systems in which more than two units are required are beyond the capability of conventional queueing techniques. Commonly, units that are processed move directly to the processor or are delayed because the processing unit is busy with another unit. Since the arriving unit must wait pending availability of the processor, a waiting line or queue develops from time to time, depending on the processor rate and the arrival rate of units to be processed. If the idea of a permanent queue associated with the processor is established, the queue at any time is either empty or contains a number of units that are delayed pending availability of the processor.

The arrival of trucks at the front end loader to be loaded with earth is a classical example of a queueing situation in construction (see Fig. 1.10). In this case, the processor (the loader or shovel) maintains its position and the trucks or haulers cycle in and out of the system. The queueing system is defined as shown in Figure

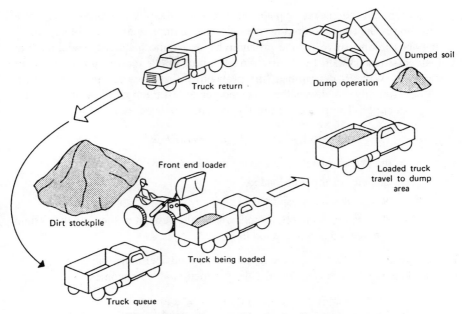

Figure 1.10 Front-end loader loading trucks.

1.11. It consists of the processor activity (loading) and the delay position (the truck queue). Units enter the system from outside its borders, are processed, and exit. In the earth-loading system, the truck exits, travels to a dump (fill) location, releases the material, and returns to the system to be reloaded. The system is not, however, directly concerned with activities that occur outside its boundaries. The "back-cycle" activities of the truck (i.e., travel to dump, dump, return to load) are characterized for the model in terms of the arrival rate of trucks at its input side. The basis (i.e., back-cycle activities) for this arrival rate is not of interest to the model. Only result in the form of the arrival rate is important.

In construction, the processor sometimes moves from point to point to serve the units to be processed. For instance, a laborer supplying bricks to masons moves from mason to mason. In this case, the masons can be thought of as the processed units and the laborer as the processor performing the processing activity of resupplying bricks. The arrival times of the masons to be processed is the rate at which

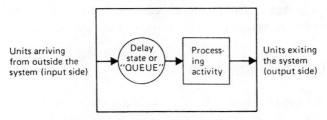

Figure 1.11 Simple queueing model.

they run through the packet or pallet of bricks supplied by the laborer. The processor time is the time required by the laborer to complete one resupply cycle.

It is possible to solve for the productivity of a finite queueing model such as the shovel–truck system by determining the probability that no units are in the system, P_0. With P_0 determined, the probability that units are in the system is $(1 - P_0)$, and this establishes the expected percent of the time the system is busy (i.e., productive). The production of the system is defined as

$$\text{Prod} = L(1 - P_0)\mu C = L(PI)\mu C$$

where μ = the processor rate (i.e., loads per hour)
  C = capacity of the unit loaded
  L = period of time considered
  $PI = (1 - P_0)$ = productivity index (i.e., the percent of the time the system contains units that are loading)

If, for instance, the PI is 0.65, the μ value is 30 loads per hour, the L value is 1.5 hr, and the hauler capacity is 15 yd^3, the production value becomes

$$\text{Prod} = 1.5(0.65)30(15) \cong 439 \text{ yd}^3$$

The value of P_0 can be determined by writing the equations of state for the system and solving for the values of P $(i = 0, M)$. The value of P_0 can be reduced to nomograph format so that, given the values of λ (the arrival rate) and μ, the production index $(PI = 1 - P_0)$ can be read directly from the chart. These nomographs for some typical queueing systems are given in Appendix B. Queueing solutions to production problems involving random arrival and processing rates are of interest, since they allow evaluation of the productivity loss caused by bunching of units as they arrive at the processor. Analysis of this effect is not possible using deterministic methods, since they consider only the interference between units caused by imbalances in the rates of interacting resources (e.g., truck and loader). This loss of productivity because of random transit times can be studied by comparing the deterministic production of a given system with the queueing production.

The queueing formulation of certain construction processes has been extended to include the concept of a storage or "hopper" in the system. This extension allows the server (i.e., loader in an earthmoving system) to store up productive effort in a buffer or storage during periods when units are not available for processing. In the context of the shovel–truck production system, this allows the loader to load into a storage hopper during periods when no trucks are available to be serviced directly. Queueing systems are discussed in detail in Chapter 3.

1.10 SIMULATION METHODS

Because of the complexity of interaction among units on the job site and in the construction environment, queueing models can be applied to only a limited num-

Figure 1.12   Two-link system.

ber of special cases. As noted in the section on line of balance, the output from one operation tends to be the input to following operations. This leads to the development of chains of extremely complex queues as well as situations in which many units are delayed at processors pending arrival of a required resource. Such chained or linked situations are too complex to be modeled using queueing models. Simulation techniques offer the only general methodology that affords a means of modeling such situations.

One class of simulation models that has been used to model chains of queues in which units interact is referred to as the "link-node" modeling format. The link-node model was developed originally to investigate problems associated with cargo handling activities at marine ports. The method gets its name from the chainlike or linked appearance of the graphical representation of the model. Nodes are located at the end of each link. Figure 1.12 is a diagram of a two-link system with the arrows indicating the direction of entity flow. Production units cycle between the nodes at either end of each link. These nodes are transfer points and normally represent operations such as loading or off-loading. Units remain in assigned links and at one end of the link can be thought of as the calling or served unit and at the other end as the server. Times or durations are associated with each node and with the half-link elements connecting each node. In the construction adaptation the nodes are designated as load and dump activities, with the upper link element representing the haul activity while the lower element representing the return travel. The nodes are constrained by the requirement for only two units (one from each link). Therefore, the ingredience requirements for the nodes in the link-node model are the same as those of the classical queueing model—a calling unit and a server. The difference is that the entity acting as the server may be delayed at the other node in its cycle awaiting realization of the ingredience requirements there. Therefore the server is not always available, and delays can occur because of the lack of a required unit. Simulation of construction operations will be the focus of most of the chapters of this text.

1.11   MODEL APPLICATION

The examples presented in this chapter indicate some of the decision problems that have been addressed by existing methods. In general, application of these models in a rigid fashion to obtain decimal-point accuracy is not relevant to the management of construction. Instead, these models and techniques provide a framework in terms of which various decision problems can be viewed. They must be adapted to the management situation. In this sense, they are analogous to the recipes used by a chef. They provide the guidelines for approaching a problem and identify the ingredients that must be blended or considered. However, too rigid an application of the recipe or model may yield a very poor result. The degree and amount of application of the model is a management decision, just as the blending of ingredients to obtain a good sauce relies on the chef's intuition. A chef must take into consideration the quality and source of the ingredients, and the manager must consider the management environment to determine the degree to which a model is relevant and can be applied.

An understanding of these and other management models, however, is very helpful in arriving at an approach to a given problem. For this reason, the study of such models is justified and aids the manager in sizing up given situations and gaining insight into the relevant aspects of the existing or potential problems.

2 Line-of-Balance Models

2.1 LINEAR CONSTRUCTION SITES

Often construction sites have linear properties that influence the production sequence. Road construction, for instance, is worked in sections that require that a set of work processes be completed in a particular sequence before the section is completed. The individual sections can be thought of as "processing through" a series of workstations.

For example, a road job may be subdivided into 14 sections that must be completed. This type of breakdown is typically established based on centerline stationing (e.g., section 1 from station 100 to station 254.3). Each of the 14 sections must undergo the following work activities: (1) rough grading, (2) finish grading, (3) aggregate base, (4) 5-in. concrete pavement, (5) 9-in. concrete pavement, and (6) curb installation.

Each of the 14 sections can be thought of as being processed by crews and equipment representing each of the six work processes. Since the site is linear, the normal way for the work to proceed would be to start with section 1, then go on to the second section and so forth. This implies that the sections will first be rough-graded, then finish-graded, then aggregate base will be placed, and so on. A bar chart indicating this sequence of activity is shown in Figure 2.1.

The bar chart indicates that work activity overlaps such that several operations are in progress simultaneously during the middle of the job. The required sequentiality leads to a "boxcar" effect. That is, a section must complete 5-in. concrete before preceding to 9-in. concrete. Therefore, the sections can be thought of as a "train" of work that must pass each of the stations represented by the six construction processes.

Many types of projects exhibit a similar kind of rigid work sequence. A high rise building, for instance, requires each floor to pass through a set of operations. Each floor can be thought of as a "car" in the "train" of work to be completed. Construction processes such as erect formwork, install reinforcing steel and imbedments, and pour concrete can be viewed as workstations through which each floor must pass.

Tunnels are worked in sections in a fashion similar to road or pipeline work. Each section must be processed through work processes such as drill, blast, remove muck, and advance drilling shield. This again leads to a repetitive sequence that is rigidly sequenced. Line-of-balance-based models are well suited to analysis and control of these kinds of projects.

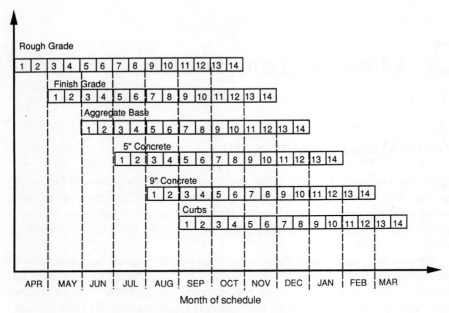

Figure 2.1 Road job bar chart.

2.2 PRODUCTION CURVES

Bar charts provide only limited information in modeling projects. They typically do not readily reflect the production rate or speed with which sections or units are being processed in linear projects. Since the rate of production will vary across time, this has a major impact on the release of work for following work processes. Delays in achieving the first units of production occur as a result of mobilization requirements. As the operation nears completion, the rate of production typically declines because of demobilization or closeout considerations. The period of maximum production is during the midperiod of the process duration. This leads to a production curve with the shape of a ''lazy S'' as shown in Figure 2.2. The slope of the curve is flat at the beginning and the end, but steep in the midsection. The slope of the curve is the production rate.

These curves are also called *time–distance* or *velocity* diagrams since they relate units of production (i.e., quantities or distance) on the *y* axis (vertical; ordinate) with time plotted on the *x* axis (horizontal; abscissa). The slope of the curve relates the increase in units of production on the *y* axis with the increment of time as shown on the *x* axis. The slope of the curve, therefore, represents the number of units produced over a given time increment. This is the rate of production.

The production curves for the road job described above are shown in Figure 2.3. The curves indicate the beginning and ending points in time for each of the processes. The slope of each curve is the production rate for each process. The distance between the beginning points of each process establishes the ''lag'' be-

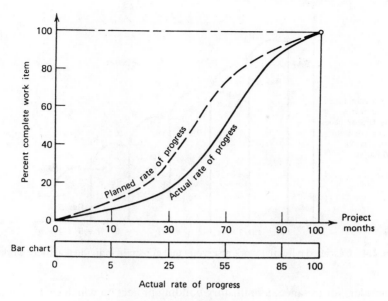

Figure 2.2 Production curve.

tween processes. The aggregate base operation begins in week 6 and lags the finish-grading operation by 2 weeks. This means that 2 weeks of work (i.e., completed finish-grade sections) are built up before the aggregate base operation is started.

Leading processes generate work area or availability so that follow-on processes have a "reservoir" of work from which to operate. This concept of work availability being generated to following processes is shown schematically in Figure 2.4. Reservoirs of work are cascaded so that units of work must be available from an "upstream" process reservoir before work is available at a lower process reservoir. This illustrates that work flow moves from leading to following processes.

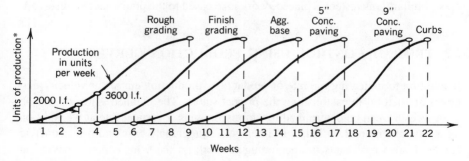

*Units of production (e.g. l.f. for rough and finsh grading, tons of aggregate base, cu. yd. or sq. ft. of paving, etc.)

Figure 2.3 Velocity diagrams for a road construction project.

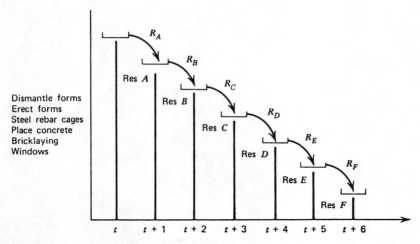

Dismantle forms
Erect forms
Steel rebar cages
Place concrete
Bricklaying
Windows

Figure 2.4 Cascaded potential work reservoirs that constrain individual activity work rates.

Consider, for instance, a multistory building project on which each floor passes through the following processes:

1. Erect forms.
2. Place steel reinforcement.
3. Place concrete.
4. Remove forms.
5. Install exterior curtain wall.
6. Install glazing.

As a floor completes one process, it enters the work reservoir of the following process. If we assume that formwork is available for four floors, once the four sets of forms have been erected, work is constrained until a form set is removed. However, the erection of forms releases work to the reinforcement–embedment process. Similarly, as resteel is placed, work is released to the pour concrete reservoir.

2.3 PROJECT CONTROL USING PRODUCTION CURVES

In addition to indicating the rate of production, production curves or velocity diagrams are helpful in establishing the project status. The planned status of the job as of week 12 can be determined by simply drawing a vertical line at week 12 on the x axis of Figure 2.3. This will intersect the aggregate base and 5-in. concrete curves. It also represents the beginning of work on the 9-in. concrete pavement (overlaying the 5-in. base concrete). It can be readily determined that

1. Both rough and finish grading should be completed.
2. Approximately 80% of the aggregate base has been placed.

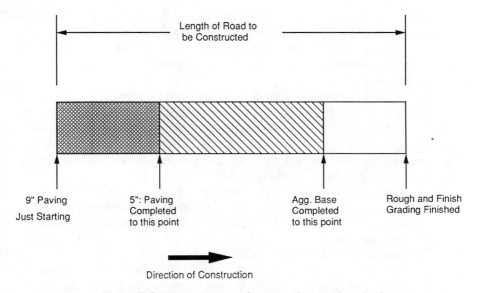

Figure 2.5   Planned status of construction as of week 12.

3. Placement of the 5-in. concrete base is approximately 30% complete.
4. Placement of 9-in. concrete is just commencing.

The planned status of construction as of week 12 is shown in Figure 2.5.

There is a definite advantage in balancing the production rates between processes. If this is not done, the situation shown in Figure 2.6 can develop. In this example, the slope (production rate) of process *B* is so steep that it catches or intersects the process *A* curve at time *M*. This requires a shutdown of process *B*

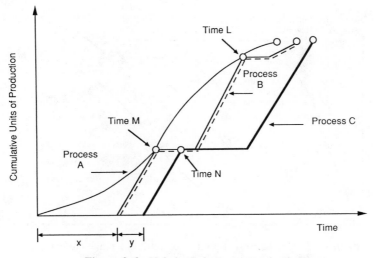

Figure 2.6   Unbalanced process production rates.

until more work units can be made available from *A*. Again, at time *L*, process *B* overtakes the production in process *A*, resulting in a work stoppage. This is clearly inefficient since it requires the demobilization and restarting of process *B*. The stoppage of process *B* at *M* also causes a "ripple" effect since this causes a shutdown of process *C* at time *N*.

It should be clear that these stoppages are undesirable. Thus, processes should be coordinated so as to avoid intersections of production curves (e.g., times *M*, *N*, and *L*). Obviously, one way to avoid this is to control production in each process so that the slopes of the curves are parallel. This implies the need to design each process so that the resources utilized result in production rates that are roughly the same for all interacting construction processes. Since the six curves for the road job are roughly parallel, we can assume that the production rates have been coordinated to avoid one process overtaking its leading or preceding process.

Production curves also help reveal poorly coordinated processes. Figure 2.7 shows the three processes involved in a tunneling operation: (1) drill and blast, (2) rock bolt installation, and (3) shotcreting.

One of the curves starting at the origin indicates the rate of production using a new advanced tunneling system. The other line starting at the origin indicates the rate of production using the existing plant. High cost would be involved in purchasing the new equipment. This is clearly not justified, however, since the slower rates of rock bolting and shotcreting control any advantage to be achieved with the new equipment. In fact, the advanced system would have to be stopped to wait for the rock bolting to catch up before proceeding. This can be compared to a race in which teams of three runners are competing. It is the time of the slowest runner that controls the finish time for a given team. Therefore, if a very fast runner (runner 1) is teamed with two relatively slower runners (runners 2 and 3), runner 1's speed will be lost since runner 3 will control. Ideally, the runners should be matched to run at roughly the same speed.

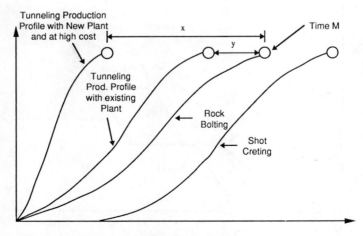

Figure 2.7 Comparison of production rates in a tunneling context.

The need for coordinated or balanced production rates between interacting processes should be clear from this discussion.

2.4 LINE-OF-BALANCE CONCEPTS

Line of balance (LOB) is a graphical method for production control integrating barcharting and production curve concepts. It focuses on the planned versus actual progress for individual activities and provides a visual display depicting differences between the two. Indication of these discrepancies enables management to provide accurate control in determining priorities for reallocation of labor resources. Those activities indicated ahead of schedule can be slowed by directing part or all of their labor crews to individual activities that lag behind schedule. This obviously assumes that resources are interchangeable. This can present a limitation to the application of this procedure in construction.

The LOB method serves two fundamental purposes. The first is to control production and the second is to act as a project management aid. Each of these objectives are interrelated through development and analysis of four LOB elements. These elements provide the basis for progress study on critical operations throughout the project duration. The four elements are

1. The objective chart
2. The program chart
3. The progress chart
4. The comparison

The *objective chart* is a straight-line curve showing cumulative end products to be produced over a calendar time period. The number of end products may be specified in the contract. Assume that the units being considered in this example are precast panels for the exterior of a high-rise building. A typical objective chart is shown in Figure 2.8.

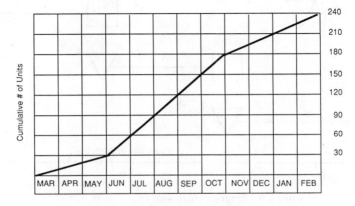

Figure 2.8 Objective chart.

This example indicates a total of 30 units to be delivered or completed by June 1, 60 units to be delivered and completed by July 1, 180 units to be delivered and completed by November 1, and a total of 240 units to be delivered and completed by February 28. The contract award date is shown as March 1.

The program chart is the basic unit of the LOB system. It is a flow process chart of all major activities, illustrating their planned, sequenced interrelationships on a "lead-time" basis. Three aspects to consider in development of the program chart are determination of (1) operations to be performed, (2) the sequence of operations, and (3) processing and assembly lead time.

The program chart indicated in Figure 2.9 describes the production process for the 240 units mentioned in the objective chart. Each activity (*A* through *E*) has associated with it a lead time (latest start time) signified by an event starting symbol (□) and an event coordination symbol (△) signifying its end or completion. These event coordination symbols, referred to as *progress monitoring points*, are labeled from top to bottom and from left to right. All five activities must be completed before one unit can be ready for delivery. This takes 30 working days as shown on the program chart's lead-time scale.

The progress chart is drawn to the same vertical scale as the objective chart and has a horizontal axis corresponding to the progress monitoring points labeled in chronologic order. Vertical bars represent the cumulative progress or status of actual performance at each monitoring point, usually based on visiting the site and measuring actual progress (e.g., assessing status of completion).

The progress chart of Figure 2.10 indicates that on a given day when inventory was taken, 120 units had passed through monitoring point 5. In other words, the vertical height of bar 5 is equal to the number of units actually completed. This corresponds to activity *E* in the program chart, which is the last activity in the

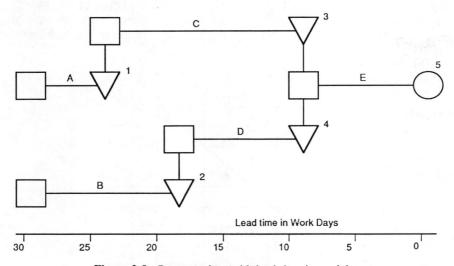

Figure 2.9 Program chart with lead time in workdays.

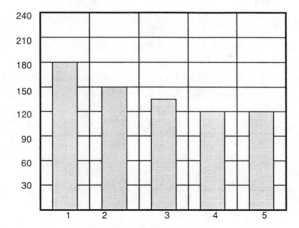

Figure 2.10 Progress chart.

production process. Similarly, activity *D* (bar 4) had completed 120 units and activity *C* (bar 3) had completed 130 units; activity *B* (bar 2) had completed 150 units; and activity *A* (bar 1) had completed 180 units.

In the *comparison*, accumulation of data ends with completion of the objective, program, and progress charts. These three charts are then utilized to draw the LOB by projecting certain points from the objective chart to the progress chart. This results in a step-down line graph indicating the number of units that must be available at each monitoring point for progress to remain consistent with the objective. Figure 2.11 indicates the LOB and the method used to project it from the objective chart to the progress chart.

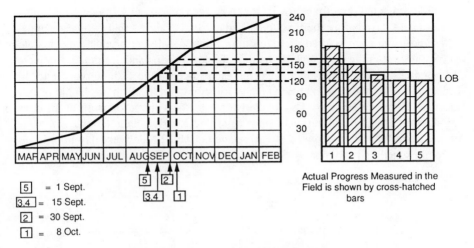

Figure 2.11 Progress chart with line of balance.

The procedure for striking the line of balance is as follows:

1. Plot the balance quantity for each control point.
   a. Starting with the study date on the horizontal axis of the cumulative delivery (objective) chart, mark off to the right the number of working days (or weeks or months, as appropriate) of lead time for that control point. This information is obtained from the program chart.
   b. Draw a vertical line from that point on the horizontal axis to the cumulative objective curve.
   c. From that point draw a horizontal line to the corresponding bar on the progress chart. This is the balance quantity for that bar.
2. Join the balance quantities to form one stair-step-type line across the progress chart.

Analysis of the LOB reveals that activities 2 and 5 are right on schedule while activities 3 and 4 show deficit units. Activity 1 shows surplus. This surplus is the difference between the 180 units actually completed by activity 1 and the 157 units indicated as necessary by the LOB. On the other hand, activities 3 and 4 are lagging by 5 and 15 units, respectively. The LOB display enables management to begin corrective action on activities 3 and 4 to ensure that they do not impede the progress rate of the remaining units.

2.5 LOB APPLIED TO CONSTRUCTION

Suppose your requirements are to produce precast concrete beams for a project. Your production schedule consists of six distinct production stages and one supply stage (see Fig. 2.12).

The required delivery schedule is indicated in Figure 2.13 with the production schedule projected directly below it. It can be seen that stage 1 (preparation) must be started 14 days before the first delivery date. It also shows that two beams must enter and leave each stage on a given day. Simply slide the production schedule to the right, aligning stage 7 with the day chosen. For instance, Table 2.1 depicts the number of beams entering and exiting each stage for the 12th delivery date. In order to have a total of 5200 beams ready for supply on the 15th day, 5200 beams must have entered stage 6. This is equivalent to the volume of work at 15 days (12 days plus the 3-day duration of stage 6). Table 2.1 gives a cumulative summary of work that must be completed at each stage on the 12th delivery day in order to maintain scheduled progress.

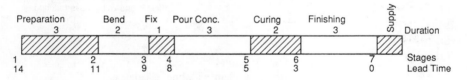

Figure 2.12 Production schedule for precast beams (Khisty 1970).

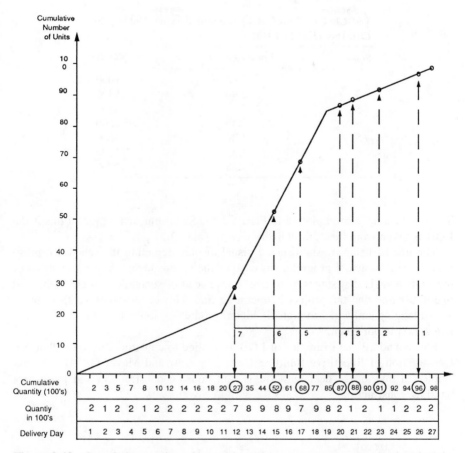

Figure 2.13 Cumulative number of beams in each stage necessary to maintain schedule.

By sliding the production schedule horizontally one day at a time from the first day of preparation (14 days before the first delivery date) to the position shown in Fig. 2.13), it is possible to develop an expected quantity table as shown in Table 2.2. This is a tabulation of the cumulative number of beams having entered any stage. As drawn, the table goes only to the 12th delivery day.

Once we have the data from Table 2.1, we are prepared to graphically display the LOB. However, before this can be compared with actual performance, a check of available progress reports must be made to determine the actual number of beams in each stage as of the 12th day. For our purpose, the following was determined for the 12th day:

Stage	1	2	3	4	5	6	7
Actual cumulative number in hundreds	88	85	83	81	76	70	30

TABLE 2.1   Cumulative Scheduled Work Volume for 12th Day (Khisty 1970)

Stage	Equivalent Day	Number
1	26th	9600
2	23rd	9100
3	21st	8800
4	20th	8800
5	17th	6800
6	15th	5200
7	12th	2700

These values have been plotted in Figure 2.14. Superimposed on top of this is the LOB as taken from the 12th delivery day of Table 2.2.

The line of balance establishes a visual display depicting the scheduled progress, as well as actual progress, for each of the seven stages. A quick analysis of the LOB reveals that stages 5, 6, and 7 are ahead of schedule, which might lead one to assume that the project is operating fine. However, stages 1, 2, 3, and 4 are all lagging and will undoubtedly hinder production in the near future if corrective action is not taken.

For two additional examples of LOB as applied to construction, see "Planning Construction of Repetitive Building Units" by Carr and Meyer (1974). The examples in this article deal with single unit housing construction and multistory building construction.

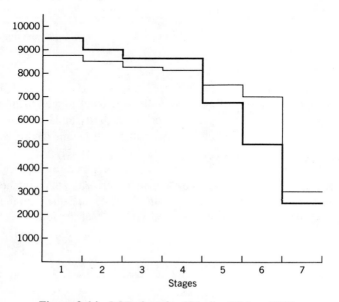

Figure 2.14   LOB chart for 12th day (Khisty 1970).

TABLE 2.2   Relationship Between Time and Number of Beams/Stage (Khisty 1970)

Day	Cumulative Quantity in Each Stage (Hundreds)						
	1	2	3	4	5	6	7
−14	2						
−13	3						
−12	5						
−11	7	2					
−10	8	3					
−9	10	5	2				
−8	12	7	3	2			
−7	14	8	5	3			
−6	16	10	7	5			
−5	18	12	8	7	2		
−4	20	14	10	8	3		
−3	27	16	12	10	5	2	
−2	35	18	14	12	7	3	
−1	44	20	16	14	8	5	
1	52	27	18	16	10	7	2
2	61	35	20	18	12	8	3
3	68	44	27	20	14	10	5
4	77	52	35	27	16	12	7
5	85	61	44	35	18	14	8
6	87	68	52	44	20	16	10
7	88	77	61	52	27	18	12
8	90	85	68	61	35	20	14
9	91	87	77	68	44	27	16
10	92	88	85	77	52	35	18
11	94	90	87	85	61	44	20
12	96	91	88	87	68	52	27

Line of balance can be as broad or narrow in detail as required. For instance, the construction example just discussed is far more specific than some LOB applications. In fact, as the complexity of the project decreases, LOB concepts become easier to use and understand.

2.6 A HIGH-RISE BUILDING EXAMPLE

To illustrate this point, consider a high-rise building in which repetitive activity sequences are a part of the floor-to-floor operation. In order to ensure a smooth flow of production, a schedule would be necessary that accounts for the interrelationships of many different activities. This becomes even more obvious when an additional constraint such as limited formwork is involved.

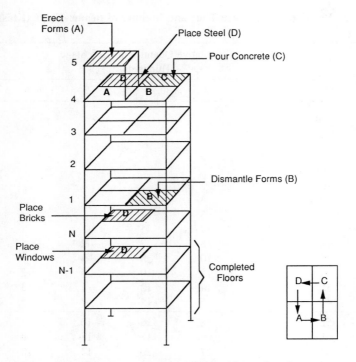

Figure 2.15   Schematic of floor cycle work tasks.

Figure 2.15 shows a schematic of the status of activities at a given point in time. The floors are divided into four sections labeled A, B, C, and D. Each of these sections can be considered as units to be processed. At the time illustrated, work is proceeding as follows:

1. Erect forms section A, floor $N + 5$.
2. Place reinforcing steel, section D, floor $N + 4$.
3. Place concrete section C, floor $N + 4$.
4. Dismantle forms section B, floor $N + 1$.
5. Place curtain wall section D, floor N.
6. Place windows section D, floor $N - 1$.

Crews proceed from section A to B to C to D.

The diagram in Figure 2.16 shows the LOB objective chart for a 10-story building. The program chart for a typical section is shown above the objective.

During the first 2 weeks the floor cycle required is one floor (four sections) per week. For weeks 2 through 6 the rate of floor production is 1.5 floors (six sections) per week. That is, six floors must be completed in the 4-week period from week 2 to week 6. In the last 2 weeks, the rate is reduced to one floor per week. The lead times required for various activities are shown on the bar program chart above

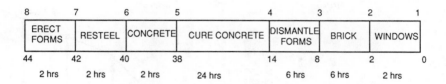

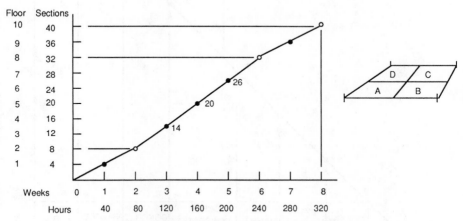

Figure 2.16 Program chart and objective.

the objective. To strike a line of balance for the study date of week 5, the lead times are projected as described in Section 2.4. A diagram of this projection is shown in Figure 2.17. The LOB values can be calculated by determining the slope relating horizontal distance (lead time) to vertical distance (required sections). The slope of the objective during weeks 5–6 is six sections (1.5 floors) per 40 hr (one week) or 6/40 sections per hour.

During the remaining weeks, the slope is four sections (one floor) per 40 hr or 1/10 section per hour. The LOB for control point I is given as

$$LOB(I) = \text{Section completed as for week 5} + (\text{slope})$$

$$\times (\text{lead time of control point I})$$

The number of sections to be completed as of week 5 is 6.5 or 26 sections (6.5 × 4). Therefore

$$LOB(1) = 26$$

$$LOB(2) = 26 + (6/40)2 = 26.3$$

$$LOB(3) = 26 + (6/40)8 = 27.2$$

$$LOB(4) = 26 + (6/40)14 = 28.1$$

$$LOB(5) = 26 + (6/40)38 = 31.7$$

$$LOB(6) = 26 + (6/40)40 = 32$$

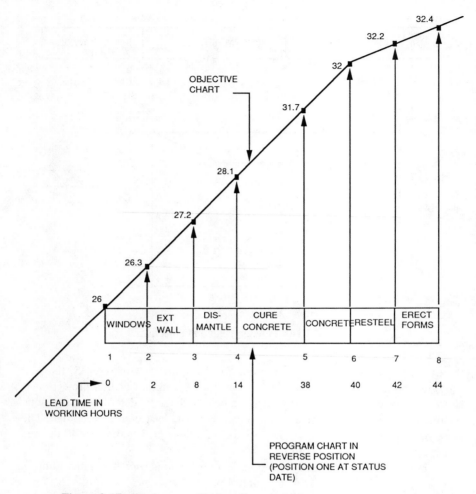

Figure 2.17 Blown-up projection of program chart onto objective chart.

Control points 7 and 8 plot to the flatter portion of the objective:

$$LOB(7) = 32 + (1/10)(42 - 40) = 32.2$$

$$LOB(8) = 32 + (1/10)(44 - 40) = 32.4$$

The line of balance for week 5 is shown in Figure 2.18. Field reports would be utilized to establish actual progress, and a comparison could be made.

2.7 DECISIONMAKING USING LOB

A slightly modified line of balance plot is shown in Figure 2.19. This LOB is based on the following assumptions, which are represented schematically in Figure 2.15:

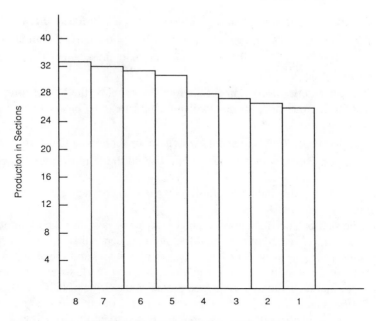

Figure 2.18   Line of balance for week 5.

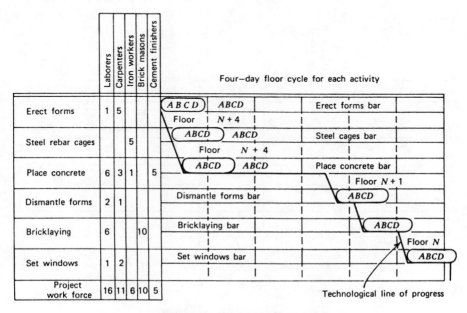

Figure 2.19   Plan for 4-day cycle.

1. Four floors of formwork and each floor divided into four sections.
2. The "set windows" process is one floor below the curtain wall activity.
3. The "install curtain wall" process is one half floor below the "dismantle forms" process.
4. The "dismantle forms" process is four floors below the "erect forms" process so that one set of forms is dismantled, it can be cleaned and reerected the following day.
5. Steel bar cages (reinforcement) are one floor section behind form erection.
6. Concreting operations are one section behind the reinforcement process.

This establishes a 4-day balanced building cycle with constant crew sizes and productivities.

In the normal balanced situation each activity crew is so matched with its work volume and relative location in the building floor sequence that as it completes one floor section (i.e., the A or B or C or D areas of a floor) and moves to the next, each of the other activity crews moves on to its next work stage. This matched performance and technological sequence produces the "line of balance" shown in Figure 2.19 as the technologic line of progress. In as current status bar chart model similar to that shown in Figure 2.1 it appears as a vertical line as shown in Figure 2.20.

If all activities proceed at a balanced rate, their completed volume of work would coincide with the line of balance. If activity rates are unbalanced, the activities will be ahead or behind schedule as indicated in Figure 2.21.

In the situation shown in Figure 2.21, the curtain wall crews follow behind (i.e., two floor sections) the dismantling crews. In Figure 2.21a the normal balanced situation is shown where, after completing the curtain wall on floor $N + 1$, the exterior finish crews can proceed to floor $N + 2$ and find clear access, since

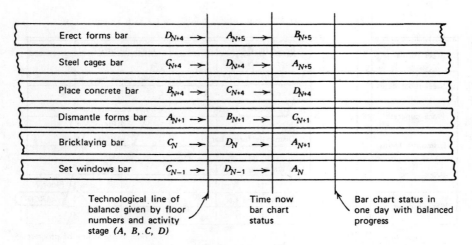

Figure 2.20   Normal LOB progress.

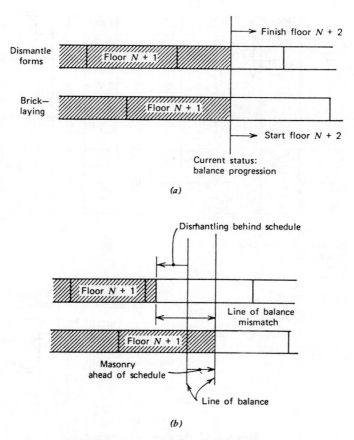

Figure 2.21 (*a*) Normal balanced situation, (*b*) out-of-balance situation (Halpin and Woodhead 1976).

the dismantling crews have half finished this floor. In Figure 2.21*b*, however, the exterior finish crews are ahead of schedule (through either better productivity or higher attendance), whereas the dismantle crews are behind schedule. In this situation the activities are out of balance, and either the exterior finish crew must be slowed down or diverted or the dismantling crews speeded up. If the dismantling crews are understaffed, then the superintendent faces the choice of

1. Working the dismantling crew longer hours to achieve more production so that the exterior finish crew can move up on schedule.
2. Increasing the strength of the crew by borrowing labor from the erection crews, which retards the form erection activity
3. Accepting the delay to the finish crew and also as a consequence on the next day a delay in form erection on the leading floor and hence on the following days delays on the steel cage and concreting crews, and so on.

Square feet	Plan	585	585	585	1070											
	Actual	552	7572													
Dismantle forms	7020	▨▨▨	↖ a'													
Square feet	Plan	585	585	585	1070											
	Actual	575	7010													
Erect forms	6435	▨▨▨	← b'													
Tons	Plan	1.83	1.83	1.83	3.66											
	Actual	1.42	19.72													
Steel rebar cages	18.30	▨▨▨	← c'													
Cubic feet	Plan	20	20													
	Actual	20	160													
Place concrete	140	▨▨▨	← d'													
Bricks	Plan	3160	3160													
	Actual	3260	3741													
Bricklaying	9480	▨▨▨	↖ e'													
Windows	Plan	4	4	4	8	4	12									
	Actual	5	13													
Windows	8	▨▨▨	f'													

a' Dismantle forms behind 6% (units working day)
b' Erect forms behind 2%
c' Steel behind 22%
d' Place concrete on schedule
e' Bricklaying 3% ahead
f' Windows 30% ahead

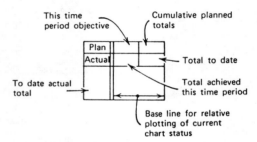

Figure 2.22 Line-of-balance chart.

The superintendent's decision will depend on the magnitude of the departures from the line-of-balance status as well as on labor availability and attitudes to the breaking up of crews and the allocation of new work assignments.

The standard line-of-balance chart is shown in Figure 2.22 together with illustrative entries relating to the status and productivity progress. The project manager updates the chart daily by entering the values of actual production on each activity. These are then used to update pictorially the bar charts to indicate current project status in relation to the planned status. This allows evaluation of the extent of departures (a', b', etc.) as a means of deciding activity priorities for project resources.

Line-of-balance concepts are intuitively followed by most building construction superintendents. The technique is sometimes used formally by building contractors on large multibuilding sites where crews can be augmented by moving personnel from job to job.

3 Queueing Systems

3.1 GENERAL CONCEPTS

In the previous chapter, units such as floor sections were processed by work processes (e.g., erect forms). The sections were considered to be discrete units arriving at workstations represented by the construction process involved. This can be thought of as a queueing situation in which the units to be processed are calling units and the construction process constitutes the server or processor.

Queueing or waiting-line situations are common to all industrial processes. As noted in Chapter 1, the arrival of trucks at a loader to be loaded with earth is a classical example of a queueing situation in construction. In this case, the processor (the loader) maintains its position and the trucks or haulers cycle in and out of the system. Another similar example is the use of dozers to push load scrapers. More common examples are ready-mix concrete trucks servicing hoppers, buggies, crane buckets, or wheelbarrows and the servicing function performed by material and man hoists on building sites.

Many questions that are of interest to the manager arise in connection with queueing problems.

1. How long will units be delayed in the queue?
2. How long will the queue be?
3. How many units can be processed by the processor, considering delays caused by queueing?
4. How can lack of service to arriving units be related to the idleness and inefficient use of processors?
5. How many processors should be provided?

These are typical of some of the problems and the types of information that the manager can utilize in designing service facilities and determining the production and resource utilization associated with such systems.

Queueing systems were first considered in a mathematical format by the Danish mathematician A. K. Erlang. In the course of studying the processing of telephone calls in Copenhagen early in this century, he was able to develop certain relationships that provide mathematically correct answers to the questions posed above. His studies provide the basis for what is now referred to as *queueing theory*. Extensions of Erlang's work have been applied to many industrial situations; more recently, certain applications to construction processes have been attempted.

Simple queueing systems can be represented schematically, as shown in Figure 1.11. They consist essentially of a processor station through which system flow units must pass. Implicit to this station is a processor unit that does the processing. This unit cycles as required back and forth from an idle to a busy state. A typical processor unit in a construction situation is a front loader, which cycles continuously from an idle state awaiting trucks to a busy state loading trucks. Obviously, if a processor (e.g., a loader) is busy with one unit, other units must wait. Therefore, a queueing model also consists of a queue or waiting line. The units that have arrived for processing, but that cannot be processed immediately, take up a waiting position in this location. This is analogous in a loader system to the truck backup or queue. Finally, the system has a border or boundary that defines when units have entered and when they have departed. On completion of processing, units are considered to have exited the system. Arrivals in the system can pass directly through the waiting position and begin processing immediately if the processor is not busy. If the processor is busy, they enter the system and are held in the queue, pending availability of the processor. The boundary of the system for incoming units is marked as the point at which they enter the queue and are either delayed or pass on directly to processing.

Units processing the system (e.g., trucks) are often referred to as *calling units*.* The processor units (e.g., the loaders) are referred to as servers. The number of servers indicates the number of channels or routes by means of which the calling units can be processed passing from the queue through a processing station and out of the system. If there is, for instance, only one loader, the system is called a single-server system. If there are several loaders, then the system becomes a multichannel or multiserver system. Units entering the system are said to enter from populations that can be either infinite or finite. In the case of a telephone system, the calls enter from a very large population and an infinite population is assumed. In an earth-loading situation, the number of trucks on the haul is finite. The number of trucks that can enter the system at any given time is known, since it is simply the number of trucks hauling minus the number in the system. For instance, in the earth-hauling model shown in Figure 3.1, six trucks are used in the system.

As shown, three trucks (indicated as T) are within the system boundaries. One truck is being loaded (i.e., processed) and two trucks are delayed in the queue. The other three trucks are outside of the system on the "back cycle." They are not within the system boundaries as defined. Therefore, the number of units that could arrive at any point in time, t, is simply the number hauling (fleet size = 6) minus the number in the system (3), or three trucks. Situations with both infinite and finite populations can be identified in construction processes.

The response of a queueing model is tied to the assumptions made regarding the rates of unit arrival, the processor rate, the type of population, and the discipline of units as they pass through the queue.

---

*This is obviously a throwback to Erlang's original work on telephone systems.

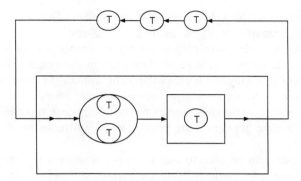

Figure 3.1 A finite system.

3.2 SYSTEM STATES

The system variables that are used to define or describe a system state (or status) at any time are called *state variables*. A particular system state is then identified by a set of instantaneous values of the state variables. If the value of one or more of the state variables is changed so that a new configuration can be recognized, then a new state exists.

Queueing theory problems can be readily described in terms of states defined as the number of units delayed in the queue, whether the processor is active or idle, and so forth. On the basis of the assumption made regarding the queueing problem model, a set of equations can be written to describe the queueing system under investigation. The concept of ''states'' is used in writing equations to describe a queueing system, and these equations are called *equations of state*. A system changes its state as time passes. In a queueing model, the configuration of the system at any given point in time can be described in terms of the number of units within the system. For instance, the finite queueing model shown in Figure 3.1 can be said to be in state S_3, since the number of units within the system boundaries is three. In this example, since the truck fleet size is six trucks, seven possible system states can be identified (i.e., when 0, 1, . . . , 6 trucks are in the system). The system can then only move between seven states as shown:

In general, the number of states in which a finite system can find itself is $M + 1$, where M is the number of calling units. For the model of Figure 3.1, the system states are S_i ($i = 0, 1, 2, . . . , 6$). The number of states characteristic of a given system is a function of the number of parameters used to define the system and the range of values associated with each parameter. In this example, one parameter (trucks in the system) was used with a range encompassing the integer values from zero to six. Had two parameters been used, the states would be given as S_{ij}. The number of states would be $N_i M_j$, where N is the number of values of i can assume and M is the number of values j can assume.

The queueing formulation of certain construction processes has been extended to include the concept of a storage or "hopper" in the system. This extension allows the server (i.e., loader in an earthmoving system) to store up productive effort in a buffer or storage during periods when units are not available for processing. In the context of the shovel–truck production system, this allows the loader to load into a storage hopper during periods when no trucks are available to be serviced directly.

Two parameters can be used to define hopper or storage-type systems. The i parameter, for example, could indicate the number of trucks in the system, and the j parameter could indicate the number of loads in the hopper or storage device. Therefore, if $i = 1, 2,$ or 3 and $j = 1$ or 2, the system has six stages, as follows:

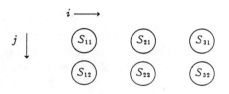

The utilization of states to define a queueing model greatly simplifies the writing of the equations that describe a particular situation.

Given a particular set of states, there is a probability P_i of being in one of those states S_i at time t and a set of probabilities T_{ij} of transiting out of one state S_i to another state S_j. Therefore, a set of transition probabilities T_{ij} exists describing the chance of moving from S_i to S_j. The graphical equivalent of this is shown in Figure 3.2. This shows the seven states of the finite system in Figure 3.1 with the states shown as circles and the transition probabilities shown as links. Only those transition probabilities from state $S_n - S_{n+1}$ or $S_n - S_{n-1}$ are considered. In other words, consider only transitions moving from $S_3 - S_4$ or $S_3 - S_2$ for state S_3. In the context of the earth-hauling situation, this means that the system can only move from containing three trucks (S_3) to containing four trucks (S_4, if a truck arrives) or two trucks (S_2, if a truck departs). In fact, it might be possible to move from S_3 to S_0, if three trucks leave the system simultaneously. Similarly, the system

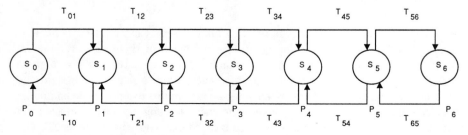

Figure 3.2   State diagram with state (P_i) and transition $T_{i,j}$ probabilities.

transits from S_1 to S_3 if two trucks arrive simultaneously. Simultaneous arrivals or departures are called "bulk" arrivals or "bulk" departures. For mathematical simplicity, it is assumed that t is defined small enough so that no more than one arrival or departure can occur in a given t. This assumption results in the diagram of Figure 3.2. State diagrams including the transition probabilities as arcs are referred to as Markovian models.

3.3 MARKOVIAN MODELS

Markovian models are helpful in representing various situations in which a system moves from state to state based on a set of transition probabilities. Howard (1960) (see Bibliography at end of this book) has presented a very clear characterization of the action of a Markovian process.

> A graphical example of a Markov process is presented by a frog in a lily pond. As time goes by, the frog jumps from one lily pad to another according to his whim of the moment. The state of the system is the number of the pad currently occupied by the frog; the state transition is of course his leap.

In particular, Markovian concepts are helpful in analyzing queueing situations.

When the graphical Markovian model is properly defined, the process of writing the equations of state for the corresponding queueing model reduces to balancing the incoming and outgoing links. At any time t, the probability of being in S_i is specified as P_i. Therefore, at t there is a probability P_n of being in state n (i.e., S_n). There are several ways that the system can be in S_n following a short interval Δt (i.e., at $t + \Delta t$) as follows.

1. The system is in S_n at t and no arrivals occur and no departures occur during Δt.
2. The system is in S_n at t and one arrival occurs and one departure occurs.
3. The system is in S_{n+1} at t and one departure occurs during Δt.
4. The system is in S_{n-1} at t and one arrival occurs during Δt.

The first two cases cover the possibility of starting in a state n (S_n) and ending in that state (S_n). That is, the transition probability of starting in S_n and remaining in S_n is

$$T_{nn} = Pr \text{ (no arrivals)} \times Pr \text{ (no services)}$$

$$+ Pr \text{ (one arrival)} \times Pr \text{ (one service)} \qquad (3.1)$$

In case 3, the transition probability of starting in a higher state $(n + 1)$ and dropping down to S_n during Δt is given as

$$T_{(n+1)n} = Pr \text{ (no arrivals)} \times Pr \text{ (one service)} \qquad (3.2)$$

Case 4 specifies the transition probability of starting in a lower state $(n - 1)$ and moving up to S_n during Δt as

$$T_{(n-1)n} = Pr \text{ (one arrival)} \times Pr \text{ (no service)} \tag{3.3}$$

In general, the probability of being in state P_n at $t + \Delta t$ is given as

$$P_n(t + \Delta t) = P_n T_{nn} + P_{n+1} T_{(n+1)n} + P_{(n-1)} T_{(n-1)n} \tag{3.4}$$

If $n = 3$, for instance,

$$P_3(t + \Delta t) = P_3 T_{33} + P_4 T_{43} + P_2 T_{23}$$

3.4 FINITE POPULATION QUEUEING MODELS

Finite population queueing models are of interest in construction, since in many situations a finite number of resources (a fleet of trucks, a crew of masons, etc.) are served by one or more resources in a cyclic fashion. This recycling of served units leads to a finite population model. For finite population systems with exponentially distributed arrival and service times, the Markovian graphical model can also be used.

The arrival transition probabilities must be modified, since the arrival rate in such systems is a function of the number of units outside of the system. That is, the rate of arrivals is proportional to the number of units that are external to the system and therefore in a mode that allows them to arrive. Returning again to the system shown in Figure 3.1, three units are outside the system and three are inside the system. The three units within the system cannot affect the arrival rate, since they are not in a mode in which they can arrive.

In this case, the arrival rate λ, which is the reciprocal of the average arrival time for one truck, $1/T_{\text{arrival(one truck)}}$,* must be multiplied by 3 to indicate the proportionality dependent on the number of trucks outside of the system. Therefore, the arrival rate of the system as shown is 3λ. If five trucks are on the back cycle, the arrival rate is 5λ. In general, the arrival rate of units entering the system is

$$(M - i) \lambda$$

where M = number of units in the finite population

$\quad i$ = number of units within the system

$\quad \lambda = 1/T_{\text{av}}$, where T_{av} is the average time a unit stays outside the system

---

*The arrival time T_{arrival}, in this situation, is equal to the average "back cycle" time of a single truck. This is the average time a unit stays outside of the system.

The arrival time T_{av}, in this situation is equal to the average back-cycle time of a single truck. This is the average time a unit stays outside of the system. A Markovian model of the six-truck system shown in Figure 3.2 is shown in Figure 3.3. Again the state probabilities, P_i, have been associated with the state circles and the arcs represent the transition probabilities between states. The arrival rates have been modified to indicate the effect of units outside the system of any state. Therefore, the probability of a unit arrival within t when the system is in S_0 is 6λ. The comparable probability of a unit arrival when in S_5 is λ. The transit probability from $S_{n+1} - S_n$ remains equal to μ. Using the method of equating inflows and outflows at each state node, $M + 1$ (e.g., seven) equations can be written. The equations written at each node in the model are as follows:

Node	Outflow	=	Inflow
0 (S_0)	$6\lambda P_0$	=	μP_1
1 (S_1)	$(5\lambda + \mu)P_1$	=	$6\lambda P_0 + \mu P_2$
2 (S_2)	$(4\lambda + \mu)P_2$	=	$5\lambda P_1 + \mu P_3$
3 (S_3)	$(3\lambda + \mu)P_3$	=	$4\lambda P_2 + \mu P_4$
4 (S_4)	$(2\lambda + \mu)P_4$	=	$3\lambda P_3 + \mu P_5$
5 (S_5)	$(\lambda + \mu)P_5$	=	$2\lambda P_4 + \mu P_6$
6 (S_6)	μP_6	=	λP_5

It is possible to solve for the productivity of a finite queueing model such as the shovel–truck system by determining the probability that no units are in the system, P_0. With P_0 determined, the probability that units are in the system is (1 − P_0), and this establishes the expected percent of the time the system is busy (i.e., productive). The production of the system is defined as

$$\text{Prod} = L(1 - P_0)\mu C = L(PI)\mu C$$

where μ = the processor rate (e.g., loads per hour)
C = capacity of the unit loaded
L = period of time considered
PI = productivity index (i.e., percent of the time the system contains units that are loading)

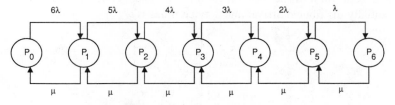

Figure 3.3 Markov model for finite system ($M = 6$).

If, for instance, the *PI* is 0.65, the μ value is 30 loads per hour, the *L* value is 1.5 hr, and the hauler capacity is 15 yd^3, the production value is

$$\text{Prod} = 1.5 \ (0.65) \ 30 \ (15) = 438.75 \ \text{yd}^3$$

The value of P_0 can be determined by writing the equations of state for the system and solving for the values of P_i ($i = 0, M$).

In addition to these equations, all state probabilities must sum to 1.0 and, therefore, the additional equation

$$\sum_{i=0}^{M} P_i = 1.0$$

is available. Since one of the node equations is redundant, this equation is substituted, providing the seventh equation required for solution of the P_i values ($i = 0, 1, 2, \ldots, 6$). Solving these equations in terms of P_0, the following values of the state probabilities result:

$$P_0 = 0.0121$$

$$P_1 = 0.0363$$

$$P_2 = 0.0906$$

$$P_3 = 0.1813$$

$$P_4 = 0.2719$$

$$P_5 = 0.2719$$

$$P_6 = 0.1359$$

The production of the system can now be calculated.

Assuming $\lambda = 6$ and $\mu = 12$, the production becomes

$$\text{Prod} = L(1 - P_0)\mu C \qquad (L = 1.5)$$

$$= 1.5(1 - 0.0121)12(15) = 266.73 \ \text{yd}^3 \ (C = 15)$$

The general form of the solution for a finite system consisting of *M* units with exponentially distributed arrival and service times is

$$P_0 = \left[\sum_{i=0}^{M} \frac{M!}{(M - i)!} \left(\frac{\lambda}{\mu} \right)^i \right]^{-1}$$

$$P_i = \left[\frac{M!}{(M - i)!} \left(\frac{\lambda}{\mu} \right)^i \right] P_0$$

The verification of the values just calculated (with the node equation) using the general form equations is left as an exercise for the reader.

The value of P_0 can be reduced to nomograph format so that, given the values of λ and μ, the production index $(PI = 1 - P_0)$ can be read directly from the chart. Such a nomograph for a system containing a single processor (e.g., loader) serving from 3 to 12 $(M = 3, 4, \ldots , 12)$ arriving units is shown in Figure 3.4. The assumed arrival and processing times are exponentially distributed. If $\mu = 12$, $\lambda = 4$, and $M = 4$, the PI value can be read on the x axis as 0.79. These nomographs for some typical queueing systems are given in Figures 3.4, 3.7, and 3.10 (after Brooks and Shaffer).

Queueing solutions to production problems involving random arrival and processing rates are of interest, since they allow evaluation of the productivity loss caused by bunching of units as they arrive at the processor. Analysis of this effect is not possible using deterministic methods, since they consider only the interference between units because of imbalances in the rates of interacting resources (e.g., truck and loader).

In addition to the production value, it may be of interest to be able to estimate the mean number of trucks in the system. This can be easily developed with the calculated state probabilities, using the formulation

$$\overline{N} = \sum_{i=0}^{M} P_i X_i$$

where $\overline{N}$ = mean number of units in the system
P_i = the probability of state i
X_i = number of units in the system state associated with P_i

Based on the information calculated above, the mean number of trucks is

X_i	P_i	$P_i X_i$
0	0.0121	0
1	0.0363	0.0363
2	0.0906	0.1812
3	0.1813	0.5439
4	0.2719	1.0876
5	0.2719	1.3596
6	0.1359	0.8154
		$\overline{N} = 4.024$

Similarly, the average queue length can be calculated using the expression

$$\overline{Q} = \sum_{i=0}^{M} P_i(X_i - 1)$$

where $\overline{Q}$ = mean number of trucks in the queue.

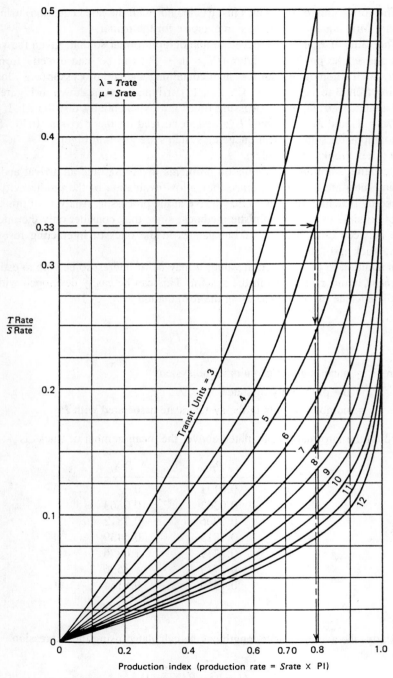

Figure 3.4 Production forecast factors (simple single-server system).

This expression reflects that in a single-server system, if one or more units are in the system at a given time, one of them is being processed. Therefore, the number of units in the queue is $(X_i - 1)$. Using this formulation, the mean number of trucks in the queue for the six-truck system is

X_i	P_i	$(X_i - 1)$	$P_i(X_i - 1)$
1	0.0363	0	0
2	0.0906	1	0.0906
3	0.1813	2	0.3626
4	0.2719	3	0.1857
5	0.2719	4	1.0876
6	0.1359	5	0.6795

$$\overline{Q} = 3.036$$

Information regarding the percent of the time that more than n units $(N < M)$ are in the system or the queue can be calculated by summing the appropriate subcomponents of the above expressions.

3.5 MULTISERVER FINITE POPULATION MODELS

In the finite system just considered, only a single-server channel was defined. If the system of Figure 3.1 is modified slightly to have two loaders, a multiserver system (number of channels = 2) is defined. A slight modification of the Markov representation allows the writing of the steady-state equations for this system. The revised model is shown in Figure 3.5.

The modification of the model relates to the transition probabilities associated with shifts from a higher to a lower state (i.e., $S_{n+1} - S_n$). Since two loaders are available, the probability of transiting down from states containing two or more units $(S_n > S_2)$ is 2μ instead of μ. This indicates that the probability of a service completion in Δt when two loaders are defined is twice that when using only one loader. Similarly, if three loaders had been defined, the probability of downshifts for states containing three or more units would be 3μ. The model for a three-server system is shown in Figure 3.6. It should be noted that the multiplier associated with μ in specifying downshifts cannot exceed the number of units in the state

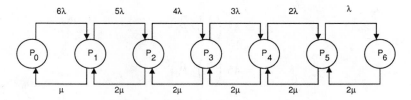

Figure 3.5 Markov model for two-server system.

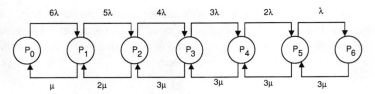

Figure 3.6 Markov model for three-server system.

from which the shift occurs. In other words, the transition probability of down-shifting from P_2 cannot be 3μ, since this implies that all three loaders are active with only two units in the system. Obviously, only two loaders can be active and, therefore, the transition probability is 2μ.

Returning to the two-server model, the equations of state for this system become

Node	Outflow	=	Inflow
0 (S_0)	$6\lambda P_0$	=	μP_1
1 (S_1)	$(5\lambda + \mu)P_1$	=	$6\lambda P_0 + 2\mu P_2$
2 (S_2)	$(4\lambda + 2\mu)P_2$	=	$5\lambda P_1 + 2\mu P_3$
3 (S_3)	$(3\lambda + 2\mu)P_3$	=	$4\lambda P_2 + 2\mu P_4$
4 (S_4)	$(2\lambda + 2\mu)P_4$	=	$3\lambda P_3 + 2\mu P_5$
5 (S_5)	$(\lambda + 2\mu)P_5$	=	$2\lambda P_4 + 2\mu P_6$
6 (S_6)	$2\mu P_6$	=	λP_5

Again, since one equation is redundant, the equation $\Sigma_{i=0}^{M} P_i = 1.0$ is used to provide the $(M + 1)$th equation required to solve for all state probabilities. The state probabilities calculated for this system by solving the set of linear equations with $\lambda = 6$ and $\mu = 12$ are

$$P_0 = 0.0622$$

$$P_1 = 6\left(\frac{\lambda}{\mu}\right) P_0 = 0.1867$$

$$P_2 = 15\left(\frac{\lambda}{\mu}\right)^2 P_0$$

$$P_3 = 30\left(\frac{\lambda}{\mu}\right)^3 P_0$$

$$P_4 = 45\left(\frac{\lambda}{\mu}\right)^4 P_0$$

$$P_5 = 45\left(\frac{\lambda}{\mu}\right)^5 P_0$$

$$P_6 = 22.5\left(\frac{\lambda}{\mu}\right)^6 P_0$$

The calculation of productivity for this system is also modified, since there are two loaders. During the periods that two or more units are in the system, the productivity is a function of 2μ. During the periods when only one unit is in the system (i.e., S_1), the production is constrained to μ. Therefore, the expression for the production of this system is

$$\text{Production} = \{\mu P_1 + 2\mu[P_2 + P_3 + P_4 + P_5 + P_6]\}CL$$

Using the same assumptions used previously regarding C and L leads to a productivity value for the two-server system of

$$\text{Production} = \{12(0.1867) + 2(12)(0.7511)\}(15)(1.5)$$

$$\approx 456 \text{ yd}^3$$

The productivity expression for three-server and multiserver systems generally must reflect the fact that the weighing factor used for μ in production calculations cannot exceed the n value of the state probability in multiples. In other words, a state that contains M units cannot yield a production of $N\mu$ if $M < N$. Therefore, the productivity expression for the three-server system becomes

$$\text{Production} = \{\mu P_1 + 2\mu P_2 + 3\mu(P_3 + P_4 + P_5 + P_6)\}CL$$

As an exercise, the reader should write the equations of state and solve for the production of the three-server system shown in Figure 3.6.

As with the single server, nomographs that give the productivity index as a function of the number of servers, number of transit units, and utilization factor (λ/μ) are available. The appropriate chart for the two-server system is shown in Figure 3.7. Entering this chart with the utilization factor ($T_{\text{rate}}/S_{\text{rate}}$) of 6/12 (0.5), the productivity index (*PI*) can be read as 1.70. System productivity is given as

$$\text{Production} = (PI)(\mu)\ CL = (1.7)(12)(15)(1.5) = 459 \text{ yd}^3$$

Methods similar to those for the single-server model can be used to calculate the mean number of units in the system and in the queue.

3.6 FINITE MODELS WITH STORAGE

Often in construction processes it is advantageous to store the productivity of the server, so that the server need not be idle while transit units are not available for service. This storage of effort allows a faster service rate when transit units do become available. The classical example of this in construction is the hopper used in earthmoving operations.* In fact, most storage towers or hoppers can be viewed

---

*Because of this fact, storage systems in construction are commonly referred to as "hopper" systems.

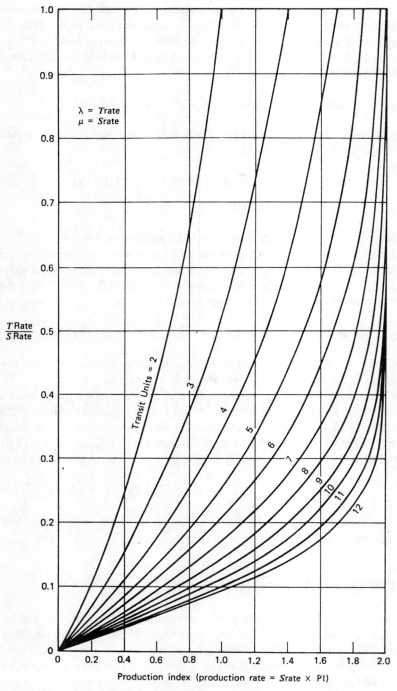

$\lambda = T\text{rate}$
$\mu = S\text{rate}$

$\frac{T\text{Rate}}{S\text{Rate}}$

Transit Units = 2
3
4
5
6
7
8
9
10
11
12

Production index (production rate = Srate × PI)

Figure 3.7 Production forecast factors (simple two-server system).

as queueing models in which the server's productivity is stored. In the earthmoving situation, a loader lifts material into a storage tower until the capacity of the tower is reached. On the arrival of a truck unit, the material is released from the tower at an accelerated rate. Similar situations arise in concrete batching and transportation as well as in masonry operations. Moreover, any construction process in which the effort of one resource is stored temporarily pending its transfer can potentially be modeled as a hopper-type situation.

Again, it is helpful, in analyzing a system with storage, to develop the system states. As mentioned previously, two parameters are used to define the state of a system that includes a hopper. The number of units in the system is still of interest and must be defined. In addition, the number of loads or services in the hopper must also be considered. Therefore, the state is defined as S_{ij}, where i is the number of transit units in the system and j represents the number of service loads in the storage. The parameter i has the range of integer values from 0 to M, where M is the number of transit units defined. The parameter j has the range of integer values from 0 to H, where H is the capacity of the storage in service loads. Some typical hopper situations are shown in Figure 3.8.

In the first instance a tower with a capacity of four truck loads is used ($H = 3$). Therefore, the range of j is (0,3). In the second example, brick packets are stored on a scaffold. The space on the scaffold for stacking brick acts as a hopper in storing the effort of the labor resupplying bricks. In this case (as shown), the space available allows storage of six packets and, therefore, the hopper capacity is six ($H = 6$).

Consider a system in which the six trucks originally defined in Figure 3.1 are serviced by one loader. This loader can either serve them directly (i.e., drop ma-

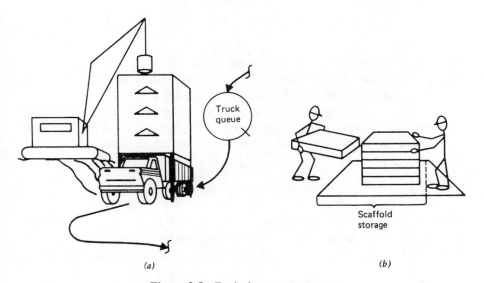

(a)   (b)

Figure 3.8 Typical storage system.

terial directly through the hopper into the truck bay), or it can store loads in a hopper tower. Assume that the capacity of the tower is two loads ($H = 2$). The symbols for arrival and server rates are λ and μ, respectively. The accelerated rate of loading afforded by the hopper is γ (gamma). The distribution of all times are described by the parameters $1/\lambda$, $1/\mu$, and $1/\gamma$, and are all assumed to be exponential. The number of states or "lily pads" between which this system transits is defined as $(H + 1)(M + 1) = 21$. Again, a Markovian model is handy in developing the state equations for this hopper system (see Fig. 3.9).

The addition of the j parameter makes this model more complex than those developed previously. The highest row of the model is identical to the single-server model shown in Figure 3.5 since, when the hopper is empty ($j = 0$), the system is actually a single-server system. However, when there are no trucks in the system ($i = 0$), the loader loads the hopper (rate $= \mu$). Therefore, the arc representing loading of the hopper at rate μ links P_{00} and P_{01}. Additionally, truck arrivals when the system is in state P_{01} cause jumps to the right down the second row of states (S_{i1}, $i = 0,1,2,\ldots,6$). If no truck arrives in a small t, the system can transit up to P_{02} following an additional load that fills the hopper to capacity. Once the system has arrived in any of the states in the middle or bottom rows, the loading of trucks is accomplished using the hopper. Therefore, the rate of loading is γ and the system transits from S_{ij}, to $S_{i-1, j-1}$, in other words, up and to the left, as shown in Figure 3.9.

The equations of state are written again by equating inflows and outflows at each node. This leads to 21 equations, one of which is redundant. The 21 state probabilities are calculated using 20 node equations and the equation summing probabilities to 1.0. The expression for production of the hopper system is given as

$$\text{Production} = \left[\mu \left(\sum_{i=1}^{M} P_{i0} \right) + \gamma \left(\sum_{j=1}^{H} \sum_{i=1}^{M} P_{ij} \right) \right] CL$$

This expression reflects the fact that production occurs by loading out of states, causing a downshift in i. Downshifts occur in the top row at rate μ. In the other rows the rate of downshift and, therefore, of production is λ.

Fortunately, nomographs are available that allow determination of the productivity index for hopper systems. These charts use the factor λ/γ and the μ/γ factor to determine the productivity index. The expression for production using these charts is

$$\text{Production} = (PI)(\gamma)CL$$

The nomograph for a system with $H = 2$ and $M = 6$ is shown in Figure 3.10. Assuming $\gamma = 24$, and the same μ and λ values used previously ($\mu = 12$, $\lambda = 6$), Figure 3.10 gives $PI = 0.5$. Therefore, system production is calculated as

$$\text{Production} = (0.5)(24)(15)(1.5) = 270 \text{ yd}^3$$

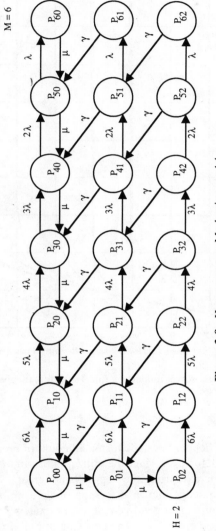

Figure 3.9 Hopper system Markovian model.

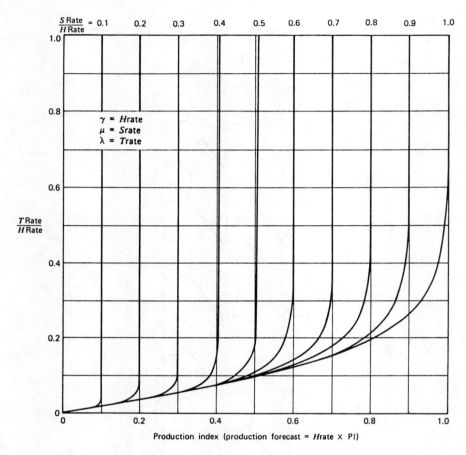

Figure 3.10 Hopper single-server production forecast factors (hopper capacity = 2, transit unit = 6).

This indicates that the hopper does add a small amount of production to the single server without hopper. However, since the utilization factor of the loader is already high, the production increase using the hopper is not great.

Although limited in scope and in the range of field problems it can handle, queueing theory does provide a good vehicle for introducing some concepts basic to modeling construction operations. The concepts of unit flows and storage, system states, delays, and processing are fundamental both to queueing systems and to the modeling of the more complex dynamics of construction processes. These concepts can be introduced by considering some basic types of queueing models presented within the content of construction processes. Having developed a framework for considering processing and delay situations, the basic elements of a more relevant construction modeling environment can be defined.

3.7 FIELD APPLICATION

To calculate production using field data, it is necessary to arrive at values for λ and μ. If a hopper system is used, a value for γ is also required. These parameters must be developed from field observations. Typical data for a single server system in which five* trucks haul to a dump location from a single loader are given in Table 3.1 (from O'Shea et al. 1964). These data indicate the sequence in which trucks arrived at the loader, the time of arrival, the time at which loading commenced, and the time of departure from the loader. As given, it is in minutes and seconds. Therefore, from the data, truck 1 arrives back in the system following its first cycle at 14 min 6 sec.

For a finite queueing situation such as this, $1/\lambda$ is the back-cycle time. This is simply the time between departure from the system and reentry. Therefore, the calculation reduces to simply

$$\overline{A} = \frac{\sum\limits_{j=1}^{M} \sum\limits_{i=1}^{N_j} (AQT_{ij} - ELT_{(i-1)j})}{\sum N_j}$$

where $\overline{A}$ = average arrival time

$\quad AQT_{ij}$ = the arrive-in-queue time for unit j on cycle i

$ELT_{(i-1)j}$ = the end-of-load time for unit j on the $(i - 1)$ cycle

$\quad\quad M$ = the number of units ($j = 1,M$)

$\quad\quad N_j$ = the number of cycles for unit j ($i = 1,N_j$)

That is, the average arrival time for a single unit is the summation for all trucks of all back cycle times divided by the total cycles for all trucks. The back cycle for an individual truck is simply the difference between the entry time for the ith cycle and the departure time for the previous cycle ($i - 1$).

To maintain the assumptions of the model, the calculation of μ must be made most carefully. Two situations are possible. If the queue is empty at the time of a unit arrival, there is no delay, and the time for commencement of loading and arrival in the system must be the same so as to maintain the assumptions of the model. The data do not always reflect this. The record of the arrival of truck 1 on its 10th cycle is the first entry in row 7 of the data. The queue at the time of arrival is empty. This can be established from the fact that the previous truck(s) departed at 150:21 (150 min 21 sec), and there is no loading until 1 arrives at 156:30. According to the assumption of the model, loading should begin being immedi-

---

*Although six trucks are indicated, truck 6 is deleted, since it cycles only twice.

TABLE 3.1 Field Data for Earth-Hauling Problem

Truck number	1	2	3	4	5	6	1	2
Arrive	0.00	0.00	0.00	9.08	9.20	9.30	14.06	16.17
Load	0.00	2.20	4.15	9.34	11.28	13.44	16.29	18.27
End of load	1.45	3.47	5.49	11.01	13.29	16.04	18.02	20.08
Truck number	3	4	5	1	3	2	6	5
Arrive	16.42	21.38	24.29	30.48	32.58	33.31	34.20	36.55
Load	20.35	22.45	26.04	31.23	34.40	37.01	39.20	42.04
End of load	22.10	25.14	28.45	33.46	36.28	38.51	41.31	44.27
Truck number	4	3	1	2	4	3	1	5
Arrive	38.47	45.55	46.28	50.07	59.00	59.37	60.55	62.43
Load	44.57	47.02	49.16	51.20	59.30	61.43	63.50	65.50
End of load	46.32	48.44	50.45	53.00	61.06	63.20	65.18	68.05
Truck number	2	4	3	1	5	2	4	3
Arrive	64.28	73.40	74.08	75.52	76.58	80.40	88.30	91.42
Load	69.21	74.19	76.54	79.20	81.34	84.40	89.02	92.50
End of load	70.56	76.22	78.44	81.03	84.07	86.45	90.27	95.18
Truck number	1	5	2	4	3	1	5	2
Arrive	92.24	93.45	96.58	106.26	107.55	110.11	110.45	115.11
Load	95.50	98.40	101.04	107.04	109.45	112.58	115.45	119.00
End of load	97.48	100.29	103.02	109.11	112.17	115.00	118.29	120.42
Truck number	4	3	5	1	2	3	4	5
Arrive	123.50	125.00	128.41	141.34	142.14	144.24	145.27	145.35
Load	125.05	128.38	132.01	141.47	143.12	145.07	146.52	148.39
End of load	127.54	131.20	135.38	142.36	144.32	146.23	148.10	150.21
Truck number	1	2	3	5	4	1	2	3
Arrive	156.30	156.43	158.52	159.41	159.57	170.18	172.36	174.56
Load	157.04	159.37	161.29	163.11	165.30	171.17	173.06	175.25
End of load	158.25	161.04	162.44	165.04	166.49	172.38	174.28	176.52
Truck number	4	1	2	3	4	1	2	3
Arrive	172.55	185.35	186.09	189.34	200.06	201.04	202.18	203.47
Load	184.25	186.17	188.10	190.00	200.52	204.00	206.07	209.57
End of load	185.49	187.37	189.29	191.19	203.36	205.27	208.30	210.47
Truck number	4	1	2	5	3			
Arrive	217.38	218.16	220.24	221.40	222.50			
Load	218.14	220.16	222.49	224.50	226.54			
End of load	219.44	221.54	224.21	226.27	228.13			

ately on arrival. However, the actual data indicate a maneuver delay such that loading does not commence until 157:04. Therefore, the processing time in this case is

$$S_{ij} = ELT_{ij} - AQT_{ij}$$

where S_{ij} = loading time of unit j on cycle i

ELT_{ij} = the end of load time for unit j on cycle i

AQT_{ij} = the arrival in queue time of unit j on cycle i

If the queue is not empty, the processing time becomes

$$S_{ij} = ELT_{ij} - ELT_{\text{preceding unit}}$$

Again, the controlling factor is not the begin load time as given in the data, since the model assumes that the begin load time of a waiting unit is assumed to occur simultaneously with the departure of its preceding unit.

To illustrate this, consider the following observations that have been converted from minutes and seconds to decimal values:

Truck Number	1	2	3
AQT	50.25	60.20	62.61
Begin load	50.75	60.55	63.62
ELT	57.80	63.15	65.35

All observations are for the sixth cycle of the truck listed. The correct model values are

$$S_{62} = 63.15 - 60.20 = 2.95 \text{ min}$$

$$S_{63} = 65.35 - 63.15 = 2.20 \text{ min}$$

This amounts to a modification of the raw field data to accommodate the assumptions of the model.

One other modification of data is required to maintain the assumptions of the model. As noted previously, the model assumes a FIFO (first-in–first-out) queue discipline. However, in the field, "breaks" in discipline sometimes occur. This happens when a unit arrives in the system but fails to process in the sequence of arrival. In effect, a later-arriving unit passes it and begins loading out of sequence, violating the FIFO criterion.

To illustrate this, consider the cycles of trucks 3 and 4 at the end of row 7 and beginning of row 8. Truck 3 arrives at 174:56 versus a 172:55 arrival time for truck 4. However, truck 3 loads at 174:56, while truck 4 does not load until 184:25. Obviously, a queue discipline break has occurred, and some changes must be made to be consistent with the basic assumptions. One approach is to change the arrival time of truck 4 so that queue discipline is not broken. The assumption is made that truck 4 arrives after truck 3. This is accomplished by modifying the *AQT* for truck 4 to a value greater than the *AQT* for truck 3. In this case, the *AQT*(4) is modified to *AQT*(3) +1 second or 174:57.

The implementation of these modifications is simplified using the worksheet shown in Table 3.2. The form allows running account of the empty–occupied status of the queue, thereby indicating which basis to use in calculating μ_{ij}. The worksheet shows observations for truck 1. Similar sheets are maintained for all trucks in the system. A summary of the calculations of λ and μ is shown in Table 3.3. Based on the calculated values of $\overline{A}$ and $\overline{S}$,

$$\rho = \frac{\overline{S}}{\overline{A}} = \frac{2.5}{13.1} = 0.19$$

Entering the nomograph with this value, the productivity index is 0.71 yielding a production value of

$$\text{Production} = (PI)\ \text{LC} = \frac{60}{2.5}\,(0.7)(1)(1)$$

$$= 16.8 \text{ truck loads per hour}$$

This is close to the actual production as reported in O'Shea et al (1964).

3.8 SHORTCOMINGS OF THE QUEUEING MODEL

This illustration of the field application of the queueing model reveals several difficulties encountered in its utilization as a precise method of forecasting production. The datum used in this example is one of 15 sets used to examine the application of these methods to forecasting shovel–truck fleet production. Two difficulties requiring correcting assumptions have already been noted: (1) breaks in queue discipline and (2) the difference between observed begin load times and the assumption of the model.

Furthermore, the assumption of exponentially distributed arrival and server times were not obtained in the field. This is indicated clearly in Figures 3.11 and 3.12, which show scaled comparisons of representative sets of observed data versus the assumed exponential distributions. Figure 3.11 shows three sets of arrival time data plotted against the assumed exponential distribution. Similarly, two sets of load time data are compared to the exponential plot in Figure 3.12. The method of scaling using λt and μt as scalers is illustrated by Table 3.4. The data as observed approximate a lognormal distribution. This is consistent with findings reported by Teicholz (1963).

In addition to these disparities between observed data and model assumptions, it is doubtful if the assumption of steady-state operation is justified. The sets of data from which the example was taken were collected during half-shift periods of 4 to 5 hour durations. Simulation studies of queueing systems similar to the example indicate that it takes a longer period to "settle down" and reach a steady-state level of operation.

TABLE 3.2 Worksheet for Finite Queue Analysis

Truck number	Trip Number	Truck Enters System	Service Completed on Truck Entering Immediately Before	Queue Yes	Queue No	Truck Exits System	Queue— Yes Service Time	Queue— No Service Time	Truck Reenters System	Back Cycle Time
①	②	③	④			⑤	⑥ = ⑤ − ④	⑦ = ⑤ − ③	⑧	⑨ = ⑧ − ⑤
1	1	0	0		X	1.75	—	1.75	14.10	12.35
1	2	14.10	16.07	X		18.03	1.96	—	30.80	12.77
1	3	30.80	28.75		X	33.77	—	2.97	46.47	12.70
1	4	46.47	48.73	X		50.75	2.02	—	60.92	10.17
1	5	60.92	63.33	X		65.30	1.97	—	75.87	10.57
1	6	75.87	78.73	X		81.05	2.32	—	92.40	11.35
1	7	92.40	95.30	X		97.80	2.50	—	110.18	12.38
1	8	110.18	112.28	X		115.00	2.72	—	141.57	26.57
1	9	141.57	135.63		X	142.60	—	1.03	156.50	13.90
1	10	156.50	150.35		X	158.42	—	1.92	170.30	11.88
1	11	170.30	166.82		X	172.63	—	2.33	185.58	12.95
1	12	185.58	185.82	X		187.62	1.80	—	201.07	13.45
1	13	201.07	203.60	X		205.45	1.85	—	218.27	12.82
1	14	218.27	219.73	X		221.90	2.17	—	Never	N/A
							$\sum = 19.31$	$\sum = 10.00$		$\sum = 173.86$

End data

$$\sum = 29.31$$

$$T_s = \sum_{i=1}^{n} tsi \div n = \frac{29.31}{14} = 0.47764 \text{ lds/min}$$

$$T_a = \sum_{i=1}^{n} tbc_i \div n = \frac{173.86}{13}$$

$$T_a = 13.3738$$

$$T_s = 2.0936 \qquad \mu = \frac{1}{T_s} = 0.47764 \text{ lds/min} \qquad \lambda = \frac{1}{T_a} = .07477 \text{ arr/min}$$

TABLE 3.3 A Summary of μ and λ Calculations

Truck	Trips	T_a	λ	Loadings	T_s	μ	Sequence
1	13	13.37	4.486	14	2.09	28.659	N/A*
2	13	12.49	4.803	14	2.34	25.618	N/A
3	14	11.86	5.056	15	2.33	25.773	T_a = 11.723 T_s = 2.359
4	12	12.97	4.624	13	2.98	20.161	T_a = 13.145
5	10	15.23	3.939	11	2.89	20.729	N/A
6	1†	34.33	N/A	2	2.63	22.817	N/A
	$\sum$ = 63	$\bar{T}_a$ = 13.07	$\bar{\lambda}$ = 4.59	$\sum$ = 69	$\bar{T}_s$ = 2.50	$\bar{\mu}$ = 24.0	

Notes. 1. Trips = Round trips. 2. λ is calculated in arrivals per hour (per truck); μ is calculated in loads per hour. 3. T_s = mean service time in minutes; T_a = mean arrival time in minutes. 4. The sequence column indicates the corrected values of T_a and T_s when breaks in the FIFO discipline occurred in the observed data.

*N/A = Not applicable.

†This row (truck 6) not included in $\bar{T}_a$ or $\bar{\lambda}$ calculations.

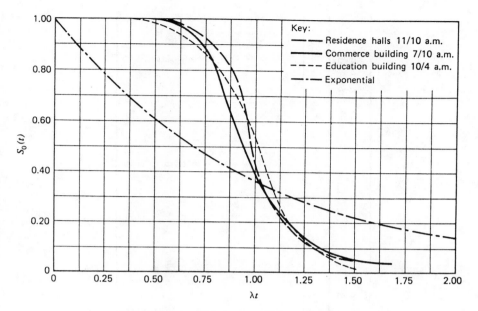

Figure 3.11 Typical actual arrival distributions.

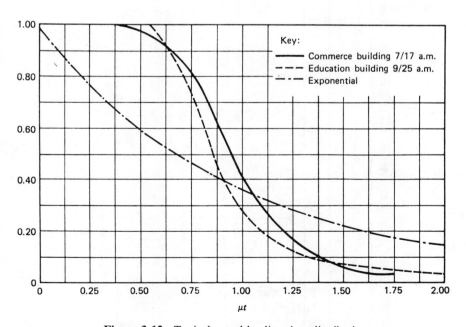

Figure 3.12 Typical actual loading time distributions.

TABLE 3.4 Typical Actual Arrival Distributions [after O'Shea et al. (1964)]

λ = 3.13 Arrivals per Hour Commerce Building 7/17 A.M.				λ = 2.27 Arrivals per Hour Residence Halls 11/10 A.M.				λ = 4.31 Arrivals per Hour Education Building 10/4 A.M.			
t_{min}	λt^*	N	$S_0(t)$	t_{min}	λt^*	N	$S_0(t)$	t_{min}	λt^*	N	$S_0(t)$
30	1.566	2	0.048	35	1.325	3	0.094	20	1.436	1	0.024
25	1.305	5	0.119	33	1.249	6	0.190	18	1.293	5	0.122
22.5	1.175	9	0.214	29	1.098	8	0.250	17	1.220	7	0.171
20.0	1.044	12	0.286	28	1.06	10	0.313	16	1.150	10	0.244
19.0	0.992	16	0.381	27	1.022	13	0.406	15.5	1.113	12	0.293
18.0	0.940	22	0.524	26	0.984	18	0.563	15.0	1.08	16	0.390
17.0	0.890	28	0.666	25	0.946	21	0.656	14.0	1.01	20	0.488
16.0	0.835	30	0.714	24	0.909	25	0.787	13.5	0.970	23	0.561
15.0	0.783	36	0.857	22	0.833	28	0.875	13	0.934	27	0.659
14.0	0.731	39	0.928	20	0.757	29	0.906	12	0.862	32	0.780
13.0	0.679	40	0.952	14	0.530	32	1.000	10	0.718	35	0.854
11.0	0.574	41	0.976					8	0.575	39	0.951
9.0	0.470	42	1.000					5	0.359	41	1.000

*λt = arrivals per hour.

Modification of the model to correct for some of the differences in field and theoretical performance can be achieved. It is possible to utilize lognormal arrival and server processes more closely approximating field observations.* It is also possible to solve the queueing model for the startup or transient phase of operation. In such solutions, dP/dt is not equal to 0 but is a function of t. However, solution of the modified model is mathematically complex and not amenable to simple nomographic presentation. It is also doubtful that the increase in precision afforded by such techniques is warranted. The relatively simple two-cycle (dual- link) interacting systems that queueing theory addresses can in most cases, be adequately handled by deterministic methods. A familiarity with the basic concepts of queueing theory is, however, helpful in understanding the dynamics of more complex systems in which chains of queues interact.

In particular, the concepts of idleness and delay, flow and activity are required to develop viable models for construction management. The discrete unit flow nature of queueing systems provides a springboard for better understanding of unit processes as they occur on the job site.

*For a more detailed discussion, see Gaarslev (1969).

4 Method Productivity Delay Model

4.1 GENERAL CONCEPTS

The *method productivity delay model* (MPDM) technique is a modification of traditional time and motion study concepts. The technique was developed to give the average construction firm a means of measuring, predicting, and improving construction method productivity (Adrian 1974). It incorporates elements of other techniques such as work sampling, production function analysis, statistical analysis, time study, and balancing models. Each of these techniques has desirable productivity measurement qualities and applications. None however, provides a convenient way of measuring, predicting, and improving productivity for the average construction firm. MPDM essentially takes a continuous sample of a construction production cycle and notes the amount and types of delays which occur during the cycle. From this data, calculations are made to determine the efficiency of the operation by showing the effect of documented delays on the productivity that is measured. From this information, the method productivity can be improved by taking appropriate action to alleviate the productivity delays. This action is based on a cost analysis that is formulated using the production increase predicted by the model when cited delays are improved.

One must at all times keep in mind that this model was developed with the average construction firm in mind. It is designed for application by a firm that cannot afford the services of professional consultants and has limited management capability available for planning and supervision. Also, it should be kept in mind that the goals of MPDM are to *measure*, *predict*, and *improve* construction method productivity. Particular emphasis is placed here because there are several methods which perform one or several of these items, but none truly perform them all.

MPDM is implemented in four stages, consisting of data collection, data processing, model structuring, and finally model implementation. The data collection stage must be preceded with the explanation of three fundamental concepts of the MPDM. These concepts establish definitions of (1) production unit, (2) production cycle, and (3) the method's leading resource.

The definition of the production unit is the basis on which the model will measure, predict, and improve method productivity. The ''production unit'' is an amount of work descriptive of the production that can easily be visually measured (Adrian and Boyer 1976). Some examples of the production unit are (1) a bucket load of concrete, (2) a truck load of transit mix poured into a concrete pump, (3) placement of a concrete block or section of blocks in a wall, and (4) placement of a section of formwork. A proper definition of the production unit is important because this determines the detail used to measure method productivity. The consequences of defining the production unit that is too small or too large in relation to other elements of the model are readily apparent.

If a concrete block is defined as the production unit, it would be practically impossible to record all the production cycles of several masons building a concrete block wall, much less note any delay information. If one defines the production unit too large, such as a complete wall poured in place by a crane–bucket pour method, the information collection would be too broad to focus on parameters that affect productivity and may cause averaging of important data.

The production cycle is simply defined as the time between consecutive occurrences of the production unit. The production cycle must be a measurable entity and be representative of productivity for each method observed. The definitions of both the production unit and the production cycle are highly correlated to the individual's experience with the method productivity being modeled.

The leading resource is the third fundamental concept. This concept is difficult to grasp until one has become familiar with the MPDM process. The leading resource is defined as the most basic or fundamental resource used in the construction method. This resource dictates the productivity of the model to the point that, if it is changed in its amount, it will change method productivity regardless of the presence or lack of current inefficiencies and regardless of the amount or makeup of the other resources. (Adrian 1974)

The concept of the leading resource is not used in the early part of the MPDM process. It is used only when there is a question about the validity of the data collected on a productivity method which has since had a resource change.

4.2 TYPES OF DELAY

After defining the production unit, cycle, and leading resource for a particular construction productivity method, one must decide what types of delay will be recognized and documented in the data collection process. Although special delay types can be defined, five types of delay are considered to be fundamental to all modern construction productivity methods. The following delays and examples of each are suggested as typical:

1. *Environment:* Change in soil conditions, change in wall section, change in roadway alignment.
2. *Equipment:* Stationary production equipment in transit, equipment operating at less-than-capable production rate.
3. *Labor:* Worker waiting for another worker, worker loafing, worker fatigue, worker not productive because of lack of knowledge of or training for the work.
4. *Material:* Material not available for equipment or labor demand, material defective.

5. *Management:* Poor planning of method resource combination and place-ment, secondary operation interfering with method productivity, poor method layout planning.

These delays apply to projects in general and can be added to or deleted from depending on the user's needs. Citing and documenting these delays in each cycle is a very skilled and experience-dependent activity. ''The model user's ability to single out productivity delay types normally increases with his practical experience and decreases as the complexity of the method in question increases'' (Adrian and Boyer 1976). The types of delay cited should be relatively independent and ob-servable while accurately defined to encompass all expected delays.

To document the delays and production cycles, the user simply and continu-ously clocks the production cycles from start to finish, citing delays or clocking them separately from the production cycle, and apportioning delays by approxi-mated percentages or actual documented ties when more than one delay occurs in a cycle. Time-lapse photography is recommended in cases where the method is very complex or the production cycle times are very short, causing the MPDM user to be unable to keep track of all the data needed.

A small example is helpful in understanding the techniques of data collection, processing, structuring, and implementation. The following example is from Ad-rian (1974). Table 4.1 is a tabulation sheet containing the collected productivity cycle and delay data. The data has been collected by the technique called *produc-tion cycle delay sampling* (PCDS). To be sure that the nondelay cycles are truly nondelay, one should always check to see that all nondelay cycle times are less than the delayed cycle times. If upon checking a nondelay cycle has greater du-ration than a delayed cycle, then obviously either a delay was not cited in the nondelay cycle or a blunder in the data collection has occurred.

The PCDS procedure is considered to be a hybrid model of time study and work sampling. Like time study, it documents actual production times in the production cycles and, like work sampling, the various states of the production method are documented. These states are either nondelayed or delayed production cycles. As for reliability of the data, the more cycles observed, the greater the confidence of the construction method productivity. The main limit here is in the economy of implementation, which is always considered critical for the average construction firm.

4.3 DATA COLLECTION

Before explaining the processing procedure of the PCDS data, it must be explained that there are two distinct methods of collecting delay information using the PCDS form. The first method is to document the length of time for each delay when it occurs within a production cycle which is also being timed. This can quickly in-crease the data collector's workload and tax his ability to see and document all delays. The other method, which is assumed to be reasonably accurate, takes the

TABLE 4.1 Sample MPDM Data (Adrian 1974)

		PRODUCTION CYCLE DELAY SAMPLING						
Page 1 of 1		Date: 7/13/81			Unit: Seconds			
Method: Contrived Example					Production unit: None			
Production Cycle (1)	Production Cycle time (sec) (2)	Environmental Delay (sec) (3)	Equipment Delay (sec) (4)	Labor Delay (sec) (5)	Material Delay (sec) (6)	Management Delay (sec) (7)	Minus Mean Non-Delay Time (8)	Remarks (9)
1	900						100	Non-Delay Cycle
2	1000		50%	50%			200	
3	750						50	Non-Delay Cycle
4	750						50	Non-Delay Cycle
5	1600					✓	800	
6	1000		✓				200	
7								
8								
9								
10								
11								
12								
13								
14								
15								
16								
17								
18								
19								
20								

mean nondelay cycle time computed in row A of the MPDM processing sheet and subtracts it from the cycle time of each cycle. This calculated value is assumed to be a reasonable approximation of the delay attributable to the delay marked in a delayed cycle. The equipment delay in the sixth production cycle in Table 4.1 is calculated as 200 sec using this approach. Here the overall cycle time was 1000 sec. If one subtracts the mean nondelay cycle time (800), 200 seconds are assumed attributable to the equipment delay. If more than one delay type is cited in a given production cycle, the resulting delay time is distributed among the delay types according to the judgment of the data collector. The calculated times allocated by the percentages are added into the "Total Added Time" row of that particular delay type on the processing form shown in Table 4.2. An example would be the equipment delay and labor delay cited in row 2 of the example problem in Table 4.1. Here, the data collector felt the delay was equally caused by both equipment and labor delays. The total delay in the cycle was computed to be 200 sec; there-

TABLE 4.2 MPDM Processing Sheet (Adrian 1974)

MPDM Processing

Date ___8/14/78___   Production Unit: ___(Seconds)___

Method: ___Example___

UNITS	TOTAL PRODUCTION TIME	NUMBER OF CYCLES	MEAN CYCLE TIME	$\Sigma(|(Cycle\ Time) - (Non\text{-}Delay\ Cycle\ Time)|)/n$
A) Non-delayed production cycles	2400	3	800	66.7
B) Overall production cycles	6000	6	1000	233.3

DELAY INFORMATION

DELAYS

	Environment	Equipment	Labor	Material	Management
C) Occurrences	0	2	1	0	1
D) Total added time	0	300	100	0	800
E) Probability of occurrence *	0	0.333	0.167	0	0.167
F) Relative severity **	0	0.15	0.10	0	0.80
G) Expected % delay time per production cycle***	0	5.0	1.7	0	13.3

* Delay cycles / total number of cycles
** Mean added cycle time / mean overall cycle time = (row D / row C) / rowB
*** Row E times Row F times 100%

fore, each delay cited gets its assigned percentage of this delay (i.e., equipment caused 100 sec of delay and labor caused 100 sec in productivity cycle 2 of the example).

4.4 DATA PROCESSING

The processing of the PCDS data consists of nothing more than adding, subtracting, multiplying, and dividing. This is in keeping with the model's attributes of ease and economy of implementation. The processing form is shown in Table 4.2 with the data from the tabulation form of Table 4.2 completely processed.

Row A: Nondelayed production cycle. This row consists of the summation of nondelayed cycle times (900 + 750 + 750 = 2400); determining the total number of nondelayed cycles (3); calculation of the mean of the nondelayed production cycles (2400/3 = 800); and a calculation of variation measure determined by subtracting the mean nondelayed cycle time (800) from each individual nondelay cycle time, summing the absolute value of these differences, and dividing by the number of nondelay cycles involved.

$$\frac{|(900 - 800)| + |(750 - 800)| + |(750 - 800)|}{3} = \frac{|100| + |-50| + |-50|}{3}$$

$$\frac{200}{3} = 66.67$$

Row B: Overall Production Cycles. This row consists of (1) the summation of all production cycles times (900 + 1000 + 750 + 750 + 1600 + 1000 = 6000), (2) the total number of overall production cycles (6), (3) the mean of the overall production cycle (6000/6 = 1000), and finally (4) a measure of variation calculated by subtracting the mean nondelay cycle time (800) from each overall production cycle time, summing the absolute value and dividing by the number of overall cycles:

$$\frac{|900 - 800| + |1000 - 800| + |750 - 800| + |750 - 800| + |1600 - 800| + |1000 - 800|}{6}$$

$$\frac{100 + 200 + 50 + 50 + 800 + 200}{6} = \frac{1400}{6} = 233.33$$

The delay information is the second segment of the processing form. In this section, the following row explanations are given:

Row C: Occurrences. This row is simply the total number of occurrences for each delay type that was documented, whether by a check or a percentage figure (i.e., equipment has two delays).

Row D: Total Added Time. The total time for each delay type is displayed here. If each delay occurrence time had been identified, one would just sum them directly, but in the example only the number of occurrences of delays was cited and not the amounts of time for each. To determine the added times when delay times have not been documented, one goes to the "Minus Mean Nondelay Time" column of the PCDS and sums up those times associated with the particular delay type always apportioning the amount by the associated percentage [equipment delay time = 50% of (200) in the second cycle and all of (200) in cycle 6].

Row E: Probability of Occurrence. This value is obtained by dividing the occurrence for each delay type by the total number of overall production cycles e.g.,

$$\left(\text{equipment} = \frac{2 \text{ occurrences}}{6 \text{ overall cycles}} = 0.333 \right).$$

Row F: Relative Severity. Row D is divided by row C to get the mean added time per occurrence for each type of delay. These values are then divided by the mean overall production time to obtain the severity rate.

[For equipment, severity rate

$$= \frac{\text{row D}}{\text{row C}} \text{ (mean cycle row B)}$$

$$= \frac{300}{2} (1000) = 0.15.]$$

Row G: Expected Percent Delay Time per Production Cycle. Row G is determined simply by multiplying row E by row F by 100% (equipment delay = $0.333 \times 0.15 \times 100\% = 5.0\%$).

4.5 MODEL STRUCTURING

With the completion of the data collection and processing elements of the MPDM, one proceeds onto the model structuring. The model structure is divided into two somewhat distinct parts. The first part uses the method productivity equation which deals with the overall or actual method production relation to ideal production as a function of the identified delays. The method productivity equation is

Overall method productivity

$$= \text{ideal productivity} (1 - E_{en} - E_{eq} - E_{la} - E_{mt} - E_{mn})$$

where E_{en} = the expected % delay due to environment/100

$\quad\quad E_{eq}$ = the expected % delay due to equipment/100

$\quad\quad E_{la}$ = the expected % delay due to labor/100

$\quad\quad E_{mt}$ = the expected % delay due to material/100

$\quad\quad E_{mn}$ = the expected % delay due to management/100

Ideal productivity is normally assumed to be the productivity measured in the non-delay production cycles. This may not always be the case because conditions may exist that indicate that some delays are not being detected (i.e., nondelay cycle durations greater than some delay cycle durations). If the user feels this to be the case, company productivity records or a detailed method study such as work sampling should be used to determine ideal productivity. It is always advisable to have a check on the ideal productivity using one of these methods.

The equation for converting nondelay cycle times of the example to ideal productivity is

$$\text{Ideal productivity} = \frac{1}{\text{mean nondelay cycle time}}$$

$$= \frac{60 \text{ min/hr } 60 \text{ sec/min}}{800 \text{ sec}} = 4.5 \text{ units/hr}$$

The rest of the right-hand side of the construction productivity equation consists of the expected % delay due to all cited delays divided by 100% and subtracted from one. This gives $(1 - E_{en} - E_{eq} - E_{la} - E_{mt} - E_{mn}) = (1 - 0 - 0.05 - 0.017 - 0 - 0.133) = 0.80$. Therefore

$$\text{Overall productivity} = (4.5 \text{ units/hr})(0.80) = 3.6 \text{ units/day}$$

To verify this overall productivity, one can calculate the overall productivity using the mean overall cycle time calculated in Table 4.2.

$$\text{Overall method productivity} = \frac{1}{\text{mean overall cycle time}}$$

$$= \frac{60 \text{ min/hr } 60 \text{ sec/min}}{1000 \text{ sec/unit}} = 3.6 \text{ units/hr}$$

NOTE: If ideal productivity is determined independently of the delay cycles, the overall method productivity can be found only by multiplication of the right-hand side of the productivity equation.)

4.6 METHOD INDICATORS

The second part of the structured MPDM deals with method indicators. There are four types of information involved here, the first of which is called the "variability of the method productivity.'' There is an ideal cycle and overall cycle variability which gives a measure of the variable nature of both cycles. The following equations are self-explanatory:

$$\text{Ideal cycle variability} = \frac{\text{Variation measure} - \text{Row A}}{\text{Mean nondelay cycle time}}$$

$$\text{Overall cycle variability} = \frac{\text{Variation measure} - \text{Row B}}{\text{Mean overall cycle time}}$$

It will be noted that this is merely dividing the last column of rows A and B by the next-to-last column of rows A and B, respectively, on the PCDS form. The results are as follows:

$$\text{Ideal cycle variability} = 66.7/800 = 0.083$$

$$\text{Overall cycle variability} = 233/1,000 = 0.233$$

Adrian (1974) states, in regard to variability, that ''Although it is difficult to set out a single set of acceptable variabilities (because of the widely-differing types of construction methods), a value greater than 1.0 for the overall cycle variability should usually be taken to mean that productivity prediction should be viewed with caution.'' The other three types of indicators are simply repeats of row E, row F, and row G of the delay information in Table 4.2. Their use is in the analysis of potential improvement. For a summary of the MPDM structure development, see Table 4.3.

4.7 USING THE METHOD RESULTS FOR IMPROVEMENT

Implementation of the MPDM is the last element to be considered here. In order to implement the MPDM for prediction and improving method productivity, one must look at the method's leading resource. The reason for this lies in the fact that if one changes a method's leading resource, then there is cause to look at the reliability and usefulness of the data, since the model is completely based on a static resource condition. The definition of a method's leading resource now becomes crucial because to improve method productivity, one must eliminate delays, and this is done by reallocating job resources. If only a supporting resource is changed, such as an extra laborer added to support a mason, then the basic model should yield a good prediction of net benefit of this laborer. This, of course, assumes that the model shows a labor delay due to the fact that the original laborers were causing a delay in method production cycles while trying to supply several masons building a concrete block wall. If we do change the method's leading resource (masons in this case), by, for instance, adding an extra mason to build the block wall, then the model user must generate a new set of data and rework the MPDM.

After considering the method's leading resource, the method productivity equa-

TABLE 4.3 MPDM Structure: Example Application

Production unit: undefined

I. *Productivity equation:*

Overall productivity = (ideal productivity) $* (1 - E_{en} - E_{eq} - E_{la} - E_{mt} - E_{mn})$
3.6 units/hr = (4.5 units/hr) $* (1 - 0 - 0.05 - 0.017 - 0 - 0.133)$

II. *Method indicators:*

A. Variability of method productivity:

Ideal cycle variability   = 67/800 = 0.084
Overall cycle variability = 233/1000 = 0.233

B. Delay information:

	Environment	Equipment	Labor	Material	Management
Probability of occurrence	0	0.333	0.167	0	0.167
Relative severity	0	0.15	0.10	0	0.80
Expected % delay time per productivity cycle	0	5.0	1.7	0	13.3

tion is looked at for a prediction of possible increases in productivity. This is accomplished by assuming that ideal productivity is constant. It is true that ideal cycle productivity varies, but this is considered negligible compared to overall cycle productivity variation (i.e., in the example 0.084 to 0.233 or about 1 to 3). The variables of the equation are overall cycle production and the E_{en}, E_{eq}, E_{la}, E_{mt}, E_{mn} factors. Therefore, if resources can be reallocated or increased in sufficient quantity or quality, then we might be able to eliminate all or part of a particular delay that is dependent on this resource and raise the overall productivity.

As an example, consider the fact that the contrived example had an equipment delay of 5%. Assume that this can be eliminated by replacing the equipment piece. Now, the overall productivity should increase from 0 to 5% over the original value. It is not assured, however, that if the equipment delay is eliminated that some other type of delay will not increase for another reason causing no gain in overall productivity. MPDM does not yield a single prediction, it only gives the user an idea of the areas in which improvements in productivity can be made based on the method productivity data collected.

Assuming the above changes of method resources are made, the MPDM method would have to be redone to verify the tentative prediction. Since resources have been changed, a reevaluation must be made. This reevaluation should serve as a better means of predicting future method productivity because it increases the sample pool from which to draw the predictions. To show the advantages of the resource change before making it, a cost analysis should be made to determine if the cost of the resource change is more than justified by the expected dollar gain in overall method productivity. Caution is urged here because predictions cannot be guaranteed due to the variabilities involved.

In summary, the MPDM method focuses attention on the measurement, prediction, and improvement of productivity in various ways. It is also relatively easy and economical to implement. It is stated that the model (1) accurately models method productivity as a function of a variable number of method resource-related parameters (delays); (2) is easy to implement because it requires little skill and does not require statistical concepts, which places it within the normal skill range of a job supervisor to use; and (3) is economical to use because it is a method that requires no computers or consultants to apply. The assumption is that a couple hours of a supervisor's or superintendent's time with a clipboard, pencil and watch to document the data and a clerk to process it are all that is needed. Further it is assumed that MPDM can be applied without interrupting the normal duties of the supervisor or superintendent applying the method.

4.8 APPLICATIONS OF MPDM

In order to evaluate the applicability of MPDM, several examples of productivity analysis using this technique will be described in the following sections. These applications indicate that MPDM has merit in short cycle situations, but is limited when longer cycle processes are encountered. This is due to the difficulties that arise during the data collection phase.

The three examples to be considered related to (1) Concrete placement with a crane and bucket, (2) placement of concrete using a concrete pump, and (3) hand mining of a soft-ground tunnel. The data collection and analysis for these examples was done by Lester M. Bradshaw, Jr. (Bradshaw, 1978).

4.9 CONCRETE PLACEMENT WITH A CRANE AND BUCKET

This example relates to the placement of concrete during the construction of the 52-story Southern Bell office building in downtown Atlanta. The process observed was the pouring of a third-story beam using the crane and bucket concrete pouring technique. The contractor used two $1\frac{1}{2}$ cubic yard concrete buckets that were loaded by a steady supply of concrete transit mix trucks. At no time were the concrete trucks not available to dump into the buckets, other than when they were maneuvering into position at the dump site. Two laborers were loading, hooking and unhooking the concrete buckets. The crane involved was a truck-mounted type that did not allow the crane operator to see the pour area three stories above. As a result, a crane signalman directed the crane operator to the bucket dump sites. There was a crew of about 10 workers dumping the concrete out of the bucket and finishing it.

The day this construction process was observed, it was overcast with occasional light rain. The inclement weather did not seem to have any effect on the construction productivity. The temperature was in the mid-80°F range since it was late July when the construction method was observed.

In order to see the complete production cycle, observations were made from the top of a four-story parking garage adjacent to the construction site where the complete cycle was in view. Before trying to apply the MPDM for the first time on this site, the crane–bucket concrete pour operation was observed on an earlier date to define the method's production unit, cycle, leading resource, and determine whether the basic five delay types (environment, equipment, labor, material, and management) would be appropriate. These five delay types were used as defined in the example application.

The production unit was defined as a 1.5-yd^3 crane–bucket load dumped at the concrete placement area along the third-story beam. It was assumed that the crane dumped equal amounts of concrete each time, even though the crane bucket had slight deviations in quantity from load to load. The error involved here was assumed to be low. The production cycle was defined simply as the time between the complete emptying of the crane bucket into the pour area and the next complete emptying of a crane bucket.

The leading resource has to be the crane, although the method is also highly dependent on the size of the crane bucket. If the size or type or number of cranes were changed, the productivity of the operation would be considerably changed. In the observed process, only one truck crane was used.

For this application, the mere existence of delays was cited because of the rapidity of production cycles and the belief that a substantial amount of nondelay cycles would be observed. The data in Table 4.4 were collected and then processed in Table 4.5.

After processing the data, the following calculations were made to determine ideal production and overall production:

$$\text{Ideal productivity} = \frac{1}{\text{mean nondelay cycle time}}$$

$$= \frac{60 \text{ min/hr } 60 \text{ sec/min}}{103.54 \text{ sec/unit}}$$

$$= 34.8 \text{ units/hr}$$

$$\text{or} = (1.5 \text{ yd}^3/\text{unit}) (34.8 \text{ units/hr})$$

$$= 52.2 \text{ yd}^3/\text{hr}$$

$$\text{Overall productivity} = \frac{1}{\text{mean overall production cycle time}} = \frac{60 \text{ min/hr } 60 \text{ sec/min}}{131.79 \text{ sec/unit}}$$

$$= 27.3 \text{ units/hr}$$

$$\text{or} = (1.5 \text{ yd}^3/\text{unit}) (27.3 \text{ units/hr})$$

$$= 40.97 \text{ yd}^3/\text{hr}$$

The MPDM structure was then developed and recorded in Table 4.6.

TABLE 4.4   Data for Crane–Bucket Concrete Process

	PRODUCTION CYCLE DELAY SAMPLING							
Page 1 of 3					Unit: First			
Method: Concrete Bucket Pour With Truck Crane					Production unit: Seconds			
Production Cycle (1)	Production Cycle time (sec) (2)	Environmental Delay (sec) (3)	Equipment Delay (sec) (4)	Labor Delay (sec) (5)	Material Delay (sec) (6)	Management Delay (sec) (7)	Minus Mean Non-Delay Time (8)	Remarks — (9)
1	157				✓		53.5	Concrete Truck change
2	121			✓			17.5	
3	158			✓			54.5	
4	134					✓	30.5	Crane operator interrupted
5	125			✓			21.5	
6	101						2.5	
7	103						0.5	
8	462				20%	80%	358.5	Delivery Truck Interference
9	97						6.5	Concrete Truck change
10	93						10.5	
11	92						11.5	
12	103						0.5	
13	97						6.5	
14	123	50%			50%		19.5	Hard Rain - Empty Truck
15	172				✓		68.5	Concrete Truck change
16	104						0.5	
17	100						3.5	
18	104						0.5	
19	97						6.5	
20	91						12.5	

From this relatively simple application of the MPDM method, it can be seen that management accounts for nearly 50% of the cited delays. These management delays were caused primarily because management did not schedule the delivery of resteel to the job site at such a time that it would not interrupt the concrete bucket pour operation. These delays could have been prevented by either scheduling deliveries at times before or after the concrete pour or making another access and crane available for the resteel delivery and off-loading. If this was done and

TABLE 4.4 (*Continued*)

PRODUCTION CYCLE DELAY SAMPLING								
Page 2 of 3					Unit: First			
Method: Concrete Bucket Pour with Truck Crane					Production unit: Seconds			
Production Cycle (1)	Production Cycle time (sec) (2)	Environmental Delay (sec) (3)	Equipment Delay (sec) (4)	Labor Delay (sec) (5)	Material Delay (sec) (6)	Management Delay (sec) (7)	Minus Mean Non-Delay Time (8)	Remarks (9)
21	99						4.5	Conc. Truck Change-No Delay
22	101						2.5	
23	106						2.5	
24	101						2.5	
25	104						0.5	
26	113						9.5	
27	130	✓					26.5	Pour Far Away & Hard
28	120						16.5	Conc. Truck Change-No Delay
29	136	✓					32.5	Pour Far Away & Hard
30	115						11.5	
31	109						5.5	
32	104						0.5	
33	112						8.5	
34	134	✓					30.5	Pour Far Away & Hard
35	111						7.5	
36	186	20%		5%	75%		82.5	Concrete Truck change
37	131	✓					27.5	Rain ends
38	144	✓					40.5	Hard corner pour area
39	121	50%		50%			17.5	Bucket misposi-tioned far away
40	121	✓					17.5	Tough corner pour

no other management delays were caused, then the expected productivity is predicted to increase from 27.3 units/hr to a revised expected overall productivity, which is computed to be:

$$\text{Overall productivity} = 34.8 \text{ units/hr} (1 - 0.043 - 0.0 - 019 - 0.047 - 0)$$

$$= 34.8 \text{ units/hr} (1 - 0.109) = 31.0 \text{ units/hr}$$

TABLE 4.4 (*Continued*)

PRODUCTION CYCLE DELAY SAMPLING								
Page 3 of 3					Unit: First			
Method: Concrete Bucket Pour With Truck Crane					Production unit: Seconds			
Production Cycle (1)	Production Cycle time (sec) (2)	Environmental Delay (sec) (3)	Equipment Delay (sec) (4)	Labor Delay (sec) (5)	Material Delay (sec) (6)	Management Delay (sec) (7)	Minus Mean Non-Delay Time (8)	Remarks (9)
41	420	10%				90%	316.5	Crane temporarily reassigned
42	109	50%			50%		5.5	Concrete Truck Change
43	106	40%				60%	2.5	Delivery Truck interference

The net productivity increase could be as much as 3.7 units/hr or 5.7 yd^3/hr. One should view this in light of the fact that this increased productivity could possibly save nearly one full hour of production time in a pour of 325 yd^3 or more. Translating this one-hour savings to such time-dependent costs as equipment rental, overhead, and labor wages indicates that a significant saving is possible. Caution must be applied here because MPDM gives only a projection of expected increased productivity and then, only, if the leading resource is not changed. This particular

TABLE 4.5 MPDM Processing Sheet for Crane–Bucket Process

MPDM Processing				
Date: ___7/28/78___		Production Unit: _(seconds) Bucket drop to_		
Method: _Concrete bucket pour with truck crane_		_bucket drop_		

UNITS	TOTAL PRODUCTION TIME	NUMBER OF CYCLES	MEAN CYCLE TIME	$\Sigma(|(\text{Cycle Time}) - (\text{Non-Delay Cycle Time})|)/n$
A) Non-delayed production cycles	2692	26	103.54	5.46
B) Overall production cycles	5667	43	131.79	32.37

DELAY INFORMATION

	DELAYS				
	Environment	Equipment	Labor	Material	Management
C) Occurrences	12	0	5	6	4
D) Total added time	245.4	0	106.4	268.7	603.7
E) Probability of occurrence *	0.279	0	0.116	0.140	0.093
F) Relative severity **	0.155	0	0.161	0.339	1.145
G) Expected % delay time per production cycle***	4.3	0	1.9	4.7	10.7

  * Delay cycles / total number of cycles
 ** Mean added cycle time / mean overall cycle time = (row D / row C) / rowB
*** Row E times Row F times 100%

process modification (e.g., alternative method of resupplying reinforcing steel) should have no influence on the leading resource (crane) and, therefore, the MPDM does not have to be updated before any projected increase in productivity is considered. It is advisable to perform a second MPDM study after implementing the proposed change to verify that the predicted crane–bucket productivity accurately

TABLE 4.6   MPDM Structure: Crane–Bucket Concrete Pour

Production unit: concrete drop

I.   *Productivity equation:*

Overall productivity = (ideal productivity) $* (1 - E_{en} - E_{eq} - E_{la} - E_{mt} - E_{mn})$
27.3 units/hr = (34.8 units/hr) $* (1 - 0.043 - 0.0 - 0.019 - 0.047 - 0.107)$
40.91 yd^3/hr = (52.2 yd^3/hr) $* (1 - 0.043 - 0.0 - 0.019 - 0.047 - 0.107)$

II.   *Method indicators:*

A.   Variability of method productivity:

Ideal cycle variability   = 5.45/103.54 = 0.053
Overall cycle variability = 32.37/131.79 = 0.246

B.   Delay information:

	Environment	Equipment	Labor	Material	Management
Probability of occurrence	0.279	0.0	0.116	0.140	0.093
Relative severity	0.155	0.0	0.161	0.339	0.145
Expected % delay time per productivity cycle	4.3	0.0	1.9	4.7	10.7

represents the method productivity. It is not clear that similar delays and productivity will result when MPDM is applied following the modification. The process may be changed in such a way that is not immediately obvious to the observer. One encouraging note is that the overall cycle variability was comparable (0.246) when compared to the example crane–bucket concrete pour overall cycle variability (0.233). This leads one to believe that an accurate measure was made of the method productivity.

4.10   PROBLEMS ENCOUNTERED IN CRANE–BUCKET APPLICATION

There are problems in applying MPDM which were encountered in this application. The first problem was in the data collection element of MPDM. It was found that a rough field PCDS form resulted as one had to document many time periods in a cycle. This problem occurred every time multiple delays happened in the production cycle defined. The time of each delay was noted, then divided by the total cycle delay time and multiplied by 100 to get the percent delays in a production cycle for each delay type. In some cycles there were as many as three and, on one occasion, four delays cited in one cycle. If only an observer's subjective opinion is used to allocate the percentage of the delay to the delay types, then the results of the process could be biased and useless for the rest of the MPDM procedure. In this study, the observer decided to only cite delays and not measure actual delay times on the final PCDS form. This process works best in rapid or short production cycles (i.e., less than 2-min cycles), except where multiple delays have to be cited during a production cycle. The vast majority of cycles were not multiple delay production cycles, so just citing a delay was appropriate. It was found later, however, that each delay *had* to be timed. The reason comes from the fact that to properly document percent splits of a delay in a multiple delay production cycle, the first time a delay occurred, its clocking had to be started. This was done because a data collector never knew when a second delay might occur in a production cycle. To illustrate, consider that a production cycle starts at 9:11:15 (9 hr 11 min 15 sec), and at 9:11:45 an equipment delay occurs and lasts until 9:12:00. The result of this is a 15-sec equipment delay. Now, if only one more delay occurs in the same cycle, from 9:12:05 to 9:12:50, then another delay has occurred of 45 seconds. By proper percentages then, the equipment delay would be $(15/60) \times 100 = 25\%$ of the total delay and the other delay would be $(45/60) \times 100 = 75\%$. Because documentation was felt to be necessary, every first delay was clocked in a production cycle. If no other delay was cited during the production cycle, then the time was ignored and only the presence of the first delay was cited. The reason for discarding this data stems from the fact that it was felt that many delays did not have a definite beginning and ending point. To circumvent this accuracy problem, one should just cite the delays in the final PCDS form but should not fall into the trap of not knowing how long the first delay in a production cycle was when the second delay occasionally appears later in the production cycle. If this happens, then subjective apportioning of delay percentages will have to be used to allocate the total cycle delay to each delay type. This will cause the data to be suspect throughout the MPDM process.

Another problem encountered in this crane-bucket concrete pour application of MPDM came in the definition of certain types of delay. The crane was located such that some parts of the pour were much closer than others. This caused the nondelay cycles to vary as much as 33%. These cycles did not have normal delays, so the modeler must either accept this as inherent in the natural cycle or credit at least some of the cycles with this delay. In this study, the observer noted that the lowest duration cycle with a delay other than an environmental delay of crane location could be used as a base figure. In this case, all cycles greater than 120 sec were considered to have an environmental delay in them if they did not contain another type of delay. Examples of this occurred in production cycles 27, 29, 34, 38, and 40 (see Table 4.4). This particular problem could be solved by exact definition of delays and explanations of what assumptions were made every time the MPDM is applied. For the average construction company, this is extremely important because consistency of the data taken must be preserved so that future use of the data is reasonable. All data taken on a particular construction process should be added to company files so that the sample size increases, yielding better predicted results.

All in all, MPDM proved a very good tool in measuring productivity on this simple process. The data collection took only 2 hr and data processing, only 1 hr. Therefore, ease and economy of implementation were realized. During data collection, however, the collector's complete attention was needed to keep up with this rapid process cycles. Based on this example, it is not feasible for a foreman or a superintendent to gather this data because of the total commitment of time and energy required for observing the process. There are several other reasons why supervisors and superintendents cannot apply this method; these will be discussed later.

The implementation element of this application of MPDM was limited to speculation since the contractor did not implement alternative access for resupply of steel. It is believed that some proper scheduling of deliveries would save substantial productivity time as pointed out in the MPDM process. This would definitely be an improvement in productivity.

Time-lapse photography is also proposed as a method of data collection when a process has extremely rapid cycles. (Adrian 1974) This may not, however, always be advantageous. An objective observer can often collect better data than a time-lapse film. Problems can occur since a time-lapse film is susceptible to vantage point limitations. A time-lapse film was taken in conjunction with data collection on this process. The results were not helpful because the camera's view did not show the entire cycle modeled. If the camera had been positioned to film all of the cycle, it would have been so far away that identifying little delays, like the crane operator being interrupted by someone talking to him, would probably not have been observed.

4.11   PLACEMENT OF CONCRETE USING A PUMP

The second application of the MPDM involved the process of concrete being pumped up to the third floor of a steel-frame high-rise and on through a header pipe and hose to the various floor slab areas being poured. This process was part of the work on the 14-story main public library in Atlanta, Georgia. The weather was sunny and the temperature was in the mid-80°F on the day the sampling process was carried out. Observations were made from an adjacent parking lot a scant 20 ft from the pouring operation. The complete cycle was readily open to a clear view at all times.

The concrete pumping operation consisted of a concrete pump truck with a pipe boom extending up to the third floor. From the boom to the slab pour area, the concrete was pumped through a combination steel pipe and flexible hose header. There were two laborers on the pump truck and at least 10 workers involved in holding the pump header hose, moving the header pipe, spreading the concrete, and finishing it. A supervisor, with a phone system, talked directly to the pump truck operator and controlled the starting and stopping of the pumped concrete.

The movement of the transit trucks was very restricted by limited access to the pump truck.

After observing the pumping procedure, the following items were defined. The only discrete production unit was the entire load of a concrete transit truck being emptied into the pump truck hopper. The production cycle was defined as the time between consecutive starts of the concrete transit truck dumping into the pump truck hopper. The concrete trucks were assumed to carry equal 9-yd^3 loads of concrete. The slight deviation among the loads of concrete was considered negligible. The leading resource for this production method was the concrete pump truck. If another pump truck were employed, the productivity would probably double. If an extra laborer were added to spread concrete, this would probably have very little effect.

Before trying to model this process, the contractor's project engineer was interviewed. From this interview, the production unit and cycle defined above were selected. Production cycles for this process were clearly of longer duration compared to the cycle time of the crane–bucket concrete pour method (27.2 min average vs. 2.2 min). Therefore, the PCDS form was reoriented to accommodate the increased number of expected delays during this much longer production cycle.

The five delay categories previously defined and used in Section 4.9 were judged to be comprehensive enough for this process. There was one innovation beyond the basic approach in the crane–bucket application. During the course of one production cycle, there were several delays under the same delay type heading (see Tables 4.7 and 4.8).

To properly document the delays, each delay duration was noted on the field PCDS form. The reason for this detail was to make the data more representative of the process and better identify possible work improvement strategies during the model implementation stage of the MPDM process. This added detail and note-taking was helpful when records were reviewed to determine how to reduce cycle delays.

Only taking one-half day's data as recommended by Adrian was judged to be inadequate for this process. The reason lies in the fact that only six production cycles were documented in about 3.5 hr of the pour. It would appear that this process should be sampled several more times and on various days to give a greater sample pool on which to base any future productivity predictions.

Nevertheless, some interesting results were obtained in the short time the process was monitored. Since the production cycles were long (average of 27.2 min), the productivity delays were all timed and documented. It is necessary to do this if the MPDM user does not have any nondelay productivity cycles or any idea of what ideal productivity should be. Therefore, by documenting each and every delay, it was a simple matter to calculate an estimated nondelay production cycle time. Since this is not normally handled on the PCDS sheet, the computed nondelay cycle time is noted in a small box in the "Minus Mean Nondelay Time" column on the PCDS form in Table 4.7. The nondelay cycle time was calculated as follows. Production cycle 1 had a total time of 28 min and a total delay time of

TABLE 4.7   MPDM Data for Concrete Pumping Process

PRODUCTION CYCLE DELAY SAMPLING

Page 1 of 1   Date: 8/2/78   Production unit: Concrete Trucks at 9 cy. each (minutes)
Method: Concrete Pumping of Floor Slabs

Production Cycle (1)	Production Cycle time (min) (2)	Environmental Delay (sec) (3)	Equipment Delay (sec) (4)	Labor Delay (sec) (5)	Material Delay (sec) (6)	Management Delay (sec) (7)	Minus Mean Non-Delay Time (8)	Remarks (9)
1	28	1			5.5	6.5	15 / 14.3 / (1.3)	Congested access for conc. trucks, Concrete truck wash-up & Ticket signing done - obstructing next truck, Header pipe change
2	30	1		1	2 / 3	10	13 / 16.3 / (0.7)	Congested access for conc. truck, Conc. pumper break, Con. slump off due to Q-wait, 2- Header pipe change, Conc. truck wash up & signing
3	21	1			1 / 3	3	13 / 7.3 / (0.7)	Congested access for conc. truck, Conc. slump off due to Q-wait, Header pipe change, Conc. truck wash up & signing
4	26	1.5			3 / 2.5 / 1	2 / 2.5	13.5 / 12.3 / (0.2)	Congested access for conc. truck, Conc. slump off due to Q-wait, Header pipe move, Conc. overpour in one slab area, Conc. truck wash-up, No conc. truck available
5	29	1.5			1 / 5	7	14.5 / 15.3 / (0.8)	Congested access for conc. truck, Conc. slump off due to Q-wait, Header pipe change, Conc. truck wash up & signing
6	29	1			5 / 4.5	2 / 3.5	13 / 15.3 / (0.7)	Congested access for conc. truck, Conc. slump off due to Q-wait, Header pipe move, Conc. truck wash up & signing
7								
8								
9								
10								

(13.0) [environment (1.0) + material (5.5) + management (6.5)]. Therefore, the nondelay cycle time for cycle 1 is (28 − 13) = 15 min and is recorded in the box. Using the assumed nondelay cycle times in the boxes of Table 4.7, an average nondelayed cycle time can be calculated as follows:

$$\text{Mean nondelayed cycle time} = \frac{15 + 13 + 13 + 13.5 + 14.5 + 13}{6}$$

$$\frac{82}{6} = 13.67 \text{ min (say, 13.7)}$$

TABLE 4.8 MPDM Processing Sheet for Concrete Pumping Process

MPDM Processing

Date: _____8/2/78_____   Production Unit:__Concrete trucks at 9 c.y.__
Method:_Concrete Pumping of Floor Slabs_   each (minutes)

UNITS	TOTAL PRODUCTION TIME	NUMBER OF CYCLES	MEAN CYCLE TIME	Σ (l(Cycle Time) - (Non-Delay Cycle Time)l)/n
A) Non-delayed production cycles	82.0	6	13.7	0.73
B) Overall production cycles	163.0	6	27.2	13.5

DELAY INFORMATION

	DELAYS				
	Environment	Equipment	Labor	Material	Management
C) Occurrences	6	0	1	6	6
D) Total added time	7.0	0	1.0	36.5	36.5
E) Probability of occurrence *	1.000	0	0.167	1.000	1.000
F) Relative severity **	0.043	0	0.037	0.224	0.224
G) Expected % delay time per production cycle***	4.3	0	0.6	22.4	22.4

* Delay cycles / total number of cycles
** Mean added cycle time / mean overall cycle time = (row D / row C) / rowB
*** Row E times Row F times 100%

To calculate the ''Minus Mean non-Delay Cycle Time'' values in the last column (before ''Remarks'') in Table 4.7, the average nondelayed cycle time is subtracted from the total cycle time observed. For the first cycle this yields

$$\text{Minus mean nondelay cycle time} = 28 - 13.7 = 14.3 \text{ min}$$

Analysis of these values for each cycle gives data for the overall cycle variability. To get information for the ideal cycle variability, the difference between the mean nondelayed cycle time and the calculated nondelayed cycle (shown in the box) are determined and placed in parentheses in the "Minus Mean Non-Delay Time" column. For the first observed cycle, this value is

$$15 - 13.7 = 1.3 \text{ min}$$

This can be seen in parentheses in the last data column in Table 4.7.

To get information for all cycles on overall cycle variability, all of the minus mean nondelay cycle time values in the last data column are averaged and appear in Table 4.8 in the last column of row B.

$$
\begin{aligned}
&\text{Summation of overall cycle variability} \\
&= \frac{14.3 + 16.3 + 7.3 + 12.3 + 15.3 + 15.3}{6} \\
&= 13.466 \text{ min (say, 13.5 min)}
\end{aligned}
$$

Similarly

$$
\text{Summation of ideal cycle variability} = \frac{1.3 + 0.7 + 0.7 + 0.2 + 0.8 + 0.7}{6}
$$

$$= 0.733 \text{ min (say, 0.73 min)}$$

These values are used later in Table 4.9 to calculate the ideal cycle variability and the overall cycle variability.

After all the data was collected, the MPDM processing was carried out and results appear in Tables 4.8 and 4.9. The following calculations were made to determine ideal productivity (assumes nondelay cycle times are representative of ideal conditions).

To verify the overall productivity, the following calculation was made and checked against the productivity equation results:

$$
\text{Overall productivity} = \frac{1}{\text{mean overall cycle time}} = \frac{60 \text{ min/hr}}{27.2 \text{ min/hr}} = 2.2 \text{ units/hr}
$$

With calculations made from the MPDM processing section of Table 4.8, the MPDM structure sheet was completed (see Table 4.9). It was quite obvious that there were large areas for work improvement in this process. The calculations show delays of 22.4% in both the material and management delay areas. With this information, it was suggested to the contractor that the following items, if corrected, could raise the productivity of the method by anywhere from 0 to 44.8%:

1. Have all concrete trucks wash up and then sign their tickets after they have moved out of the way of the next truck.

TABLE 4.9 MPDM Structure: Concrete Pumping of Floor Slabs

Production unit: concrete transit truck loads (9 yd^3)

I. *Productivity equation:*

Overall productivity = (ideal productivity) $\times$ $(1 - E_{en} - E_{eq} - E_{la} - E_{mt} - E_{mn})$
2.21 units/hr = (4.38 units/hr) $\times$ $(1 - 0.043 - 0.0 - 0.006 - 0.224 - 0.224)$

II. *Method indicators:*

A. Variability of method productivity:

Ideal cycle variability   = 0.73/13.7 = 0.053
Overall cycle variability = 13.5/27.2 = 0.50

B. Delay information:

	Environment	Equipment	Labor	Material	Management
Probability of occurrence	1.000	0	0.167	1.000	1.000
Relative severity	0.043	0	0.037	0.224	0.224
Expected % delay time per productivity cycle	4.3	0	0.6	22.4	22.4

2. Have a worker check the slump of each cement truck right before the previous concrete truck is finished dumping.
3. Try to time the movement of the header pipe with the end of each transit truck load, then the header can be moved while the next concrete truck is maneuvering to dump into the pump truck hopper.

It was felt by the contractor that only the first two suggestions could be done to improve productivity. This would entail hiring an extra person, but the contractor believed that was a reasonable expense if productivity could be improved just the 22.4% attributable to the material delays. The new overall productivity would be

Overall predicted productivity = (43.8 units/hr) $\times$ $(1 - 0.043$

$- 0.006 - 0 - 0.224) = 3.18$ units/hr or 28.7 yd^3/hr

This figure carries an increase of 0.97 units/hr or 8.7 yd^3/hr. There was no question that this would be a cost effective work improvement decision.

4.12 TUNNELING BY HAND LABOR

The third application of the MPDM was done on a very labor-intensive productivity method: tunneling by hand. The project was for Fulton County, Georgia and consisted of various diameter sewer and water lines. The general contract was for nearly one million dollars, with the tunneling portion accounting for slightly over $200,000 of that. The tunnel used for this study was 5 ft in diameter, 260 linear ft long, and constructed with steel liner plates. The only machinery involved was considered supportive to the productivity efforts of the laborers who hand-mined the tunnel. The actual construction consisted of two miners cutting out the muck with air tools and hanging the steel liner plates, two muckers disposing of the muck and supplying the miners, and the top worker who operated the crane used to dump muck cars from the tunnel. The project was being run in two shifts without a supervisor for the second shift. The weather was rainy and the temperature was in the mid-80°F the day the data were taken. The production unit was defined as the completion of one ring of steel liner plates. There was little variation in this unit. A steel liner plate ring is 16 in. long (1.333 linear ft) and 5 ft in diameter and consists of five separate plates bolted together by 60 bolts.

The production cycle was the time between successive liner plate ring completions by the miners. The application had very long cycle times involved. Many delays were expected. Also, because of the labor-intensity of this process, the labor delay type was subdivided to give a more accurate representation of the method. Another reason for the subdividing of this delay type was because the leading resource was defined to be the tunnel miners. Therefore, the labor delay was subdivided into delays attributable to miners, muckers, and top crew. The miners, it was felt, would have the greatest impact on the production if they were replaced by other miners. There was not enough space in the tunnel to increase the number of miners, but losing one of the two would also greatly affect process productivity.

The data collection for this application of MPDM was extremely difficult and highly subjective (see Table 4.10 for the data). The reasons are quite varied. For example, the entire cycle was not in clear view of the data collector. Data was collected from 200 ft up into the tunnel. This made it impossible to accurately cite the delays in the shaft where the muck car dumping and supply loading occurred. This problem was partially overcome by interviewing the mucker returning the car to the heading about any delays that occurred. An ideal dump-and-return time was assumed for the muck car and obvious deviations were considered delays for the mucker or the top crew. This was not totally accurate, but during the production cycle, this delay type occurred very rarely. Muck car dump-and-return seldom accounted for more than 20% of the actual production cycle.

Some environmental delays were based on the varying ground conditions that affected the mining time. In this particular job, the time the miners took to cut out roots and not soil was clocked as closely as possible and listed as an environmental delay. This is another rather subjective delay measure because it was hard to know what effect the roots had on regular soil-mining time, if any at all.

By far the most subjective delay was the management delay cited as lack of supervision. This was based on the observer's experience and based upon assumptions regarding reasonable work times. The delays did have some measurable value, but in general, the delays were figured by taking the experience-defined ideal time

TABLE 4.10 MPDM Data for Hand-Mining Process

PRODUCTION CYCLE DELAY SAMPLING

Page 1 of 1   Date: 8/2/78   Production unit: Completion of One Ring of Steel
Method: Tunneling by Hand Labor

Production Cycle (1)	Production Cycle time (min) (2)	Environmental Delay (sec) (3)	Equipment Delay (sec) (4)	Material Delay (sec) (6)	Management Delay (sec) (7)	Labor Delay (sec) (5)			Minus Mean Non-Delay Time (8)	Remarks (9)
						Miners	Muckers	Top Crew		
1	75	4		3	6	14	16		32 / 44.9 / (1.9)	Roots in soil; Inexperienced muckers; No bag for hay; Line & Grade Break; Ordered wrong Liner plates; Slow start; Lack of Supervision
2	70	5		6	10	10	10		29 / 39.9 / (1.1)	Roots in soil; Inexperienced muckers; No haybag or drinking water; Rail not laid when muckers were idle; Slow start; Lack of Supervision
3	40	4	3			3	2		28 / 9.9 / (2.1)	Roots in soil; Rain on top crews; Crane brakes wet; Inexperienced muckers;
4	53	5	2		7	5	3		31 / 22.9 / (0.9)	Roots in soil; Still raining; Crane breaks wet; Pumps clogging; Inexperienced muckers; No haybag or drinking water; Rail not laid when muckers were idle; Lack of Supervision
5	51	7			5	13			26 / 20.9 / (4.1)	Wet running soil; Heavy to move; Line & Grade Break; Inexperience; Lack of Supervision
6	46	4				18	4		30 / 15.9 / (0.1)	Roots & Heavy Soil; Smoke break; Rail not moved when muckers were idle; Inexperienced muckers
7	79	4			8	19	13		35 / 44.9 / (4.9)	Roots & Heavy Soil; Line & Grade break; Inexperienced muckers; Lack of supervision
8										
9										
10										

for a work activity and recording the difference between it and the actual time taken.

Since the observer in this case was the laborer's supervisor, this appeared to have a substantial but unmeasurable effect on performance. This without doubt biased the data taken. This factor cannot be ignored, yet it also cannot be taken into account.

As one can see from Table 4.11, MPDM did not lend itself to the modeling of this highly complex, labor-intensive productivity method. The results were considered interesting because they did cite in what areas delays were occurring. The values for the delays were highly subjective, but would give a cost estimator an indication of where the greatest delays occurred. It was known already that the one miner and mucker were under training at their jobs. An interesting point was that the workers pointed to poor ground conditions as the culprit for poor production. It is rather obvious from this data that there is roughly a 3.5–1 (29% versus 8%) greater delay attributable to mucker and miner delays as compared to the

TABLE 4.11   MPDM Processing Sheet for Hand-Mining Process

MPDM Processing

Date: _____7/26/78_____   Production Unit: _Completion of One Ring of Steel_
Method:_ Tunneling by Hand Labor _   Liner Plates (minutes)

UNITS	TOTAL PRODUCTION TIME	NUMBER OF CYCLES	MEAN CYCLE TIME	Σ (|(Cycle Time) - (Non-Delay Cycle Time)|)/n
A) Non-delayed production cycles	211	7	30.1	2.16
B) Overall production cycles	414	7	59.1	29.0

DELAY INFORMATION

	DELAYS						
	Environment	Equipment	Material	Management	Labor Miners	Labor Muckers	Labor Top Crew
C) Occurrences	7	5	2	2	7	6	0
D) Total added time	33.0	36.0	5.0	9.0	72.0	48.0	0
D) Probability of occurrence *	1.00	0.71	0.29	0.29	1.00	0.86	0
E) Relative severity **	0.080	0.122	0.042	0.076	0.174	0.135	0
F) Expected % delay time per production cycle***	8.0	18.7	1.2	2.2	17.4	11.6	0

* Delay cycles / total number of cycles
** Mean added cycle time / mean overall cycle time = (row D / row C) / rowB
*** Row E times Row F times 100%

environmental delays of poor soil conditions. These labor delays were expected and were due to the lack of any other qualified personnel to do the job. This problem was accepted by the contractor.

The same procedures of documenting all delays and processing the data were used as in the concrete pumping process since experienced tunneling professionals know that the occurrence of a nondelay cycle is extremely unlikely (see Table 4.12). Ideal productivity was estimated on the basis of experience and company

TABLE 4.12   MPDM Structure: Tunneling by Hand Labor

Production unit: Completion of one ring of steel liner plates

I.   *Productivity equation:*

Overall productivity = (ideal productivity) $* (1 - E_{en} - E_{mn} - E_{eq} - E_{mt} - E_{mi}$
$- E_{mu} - E_{tc})$

1.02 units/hr = (2 units/hr) $* (1 - 0.08 - 0.087 - 0.012 - 0.022 - 0.174$
$- 0.116 - 0)$

II.   *Method indicators:*

A.   Variability of method productivity:
Ideal cycle variability   = 2.16/30.1 = 0.072
Overall cycle variability = 29.0/59.1 = 0.491

B.   Delay information:

	Environ- ment	Manage- ment	Equip- ment	Material	Miners	Labor Muckers	Top Crew
Probability of occurrence	1.00	0.71	0.29	0.29	1.00	0.86	0
Relative severity	0.080	0.122	0.042	0.076	0.174	0.135	0
Expected % delay per productivity cycle	8.0	8.7	1.2	2.2	17.4	11.6	0

records to be 30 min per ring or 2 units/hr. From the documented times, the average found was

$$\frac{60 \text{ min/hr}}{30.1 \text{ min/unit}} = 1.99 \text{ units/hr}$$

Since the collected data was so subjective, this rate is of limited value.

The overall results gave a productivity rate of 1.02 units/hr or 8.16 units per 8-hr shift or 10.9 linear ft per 8-hr shift using the assumed ideal productivity. The documented average for this entire 260-linear-ft LF tunnel was 6.75 linear ft per shift. This indicates that more sampling should be carried out to give a more representative data pool. The cost estimator for this job based his bid on averages from general cost records and company history. The estimated production value was 6.0 linear ft per 8-hr shift. Therefore, MPDM must be considered with caution as a basis for cost estimates. For this particular job, MPDM would require so much data that it is doubtful whether it would be cost effective. This tunnel would be complete before a representative sample could be obtained. The real value of MPDM, in this application, was its ability to cite delays, give an approximate productivity value, and indicate where efforts might be helpful in improving process productivity.

4.13 CONCLUSIONS

In this chapter, MPDM has been reviewed and applied to several construction processes and evaluated from various perspectives. It has been found to be helpful in measuring, predicting, and improving productivity in some fairly simple construction processes. It does, however, have several limitations.

Simplified statistical measures are used with the assumption that average construction company supervisors and superintendents are not able to understand this more accepted standard. It is very doubtful whether supervisory personnel at this level have the time, ability, or motivation to apply this method. Experience in the cases cited (Bradshaw 1978) established that superintendents and foremen have difficulty applying the method even with these simplified production measures. This undermines the basic concept of MPDM, which assumes that simplified measures will make the method more accessible to relatively low-level field personnel.

Moreover, the MPDM measures are of no assistance in predicting the impact of changes of resources or technology when modifying the design of a construction process. Also, since nonstandard statistical concepts are used, comparison with other data collection schemes is difficult or impossible. From the viewpoint of building up a consistent database of production information, this is a severe limitation.

The method is very limited and of questionable value when applied to extremely short-cycle or relatively long-cycle (e.g., concrete pumping or tunneling) processes. This occurs because of value judgments that must be made by the data collector. These subjective judgments tend to undermine the objectivity of the data and impact the reliability of the results obtained.

The method is difficult to apply to complex processes with long cycle times and complex interaction of resources (multicycle systems). Processes such as concrete pumping and tunneling with many primary and secondary (sub)cycles cannot be readily analyzed using this technique.

Since the technique is based on field data collection from an ongoing process, it is not helpful as a planning and design tool prior to commencement of the work on site. This is a very significant limitation since, as in the case of the tunneling example, the project may be too short to implement the work improvement actions indicated by the MPDM analysis. Methods that aid the manager in technology and resource selection prior to work commencing in the field have a much bigger cost-saving potential. Depending on the complexity of the process, MPDM may then be used as a monitoring technique to "fine-tune" the process once it is underway in the field.

When dealing with relatively simple two-cycle systems, however, MPDM does provide a good framework for studying processes and indicating the impact of delay sources on overall production. In such cases, it is attractive as a method of troubleshooting processes in which production rates are falling below the expected level and the reason for low productivity is not apparent.

5 Process Modeling Concepts

Construction operations contain the basic work processes in construction. Their definition requires a knowledge of the construction technology involved, a breakdown of the processes into elemental work tasks, the identification of the required resources, and definition of the work assignment to the labor force involved. The description of a construction operation must indicate what is to be done and how (i.e., a technology and process focus) and who is to do it with what (i.e., a resource use focus). Practical descriptions for the performance of the construction operation must also indicate the conditions under which the various processes and work tasks can be initiated, interrupted, or terminated. The planning and management of an efficient construction operation also requires information relating to the impact on productivity and resource use of different spreads of equipment for different crew composition and sizes.

Any modeling methodology for construction operations must be capable of meeting most (if not all) of the above requirements. Construction operation models can be developed at several levels, depending on whether the model purpose is to describe, analyze, or assist the user in a decision process relating to the operation.

1. Descriptive models require simple modeling concepts if field and head office agents are to find the models useful.
2. Analysis models require the development of solution processes that operate on relevant descriptive data.
3. Decision models must focus on decision variables pertinent to the construction operation itself. The decision variables must be available to the office and field agents for manipulation in the design and management of the construction operation.

A modeling methodology that is capable of integrating a construction operation model through all these levels is preferable to a number of independent and fragmented modeling methodologies.

This chapter introduces modeling concepts concerned with description of construction operations.

5.1 A MODELING RATIONALE FOR CONSTRUCTION OPERATIONS

Construction operations can be considered and defined in terms of specific collections of work tasks, where the work task is a basic or elemental component of the work. The work task (see Fig. 5.1) is a readily identifiable component of a construction process or operation whose description to a crew member implies what is involved in and required of the work task. The various work tasks are logically related according to the technology of the construction process and the work plan. The work plan prescribes the order in which the resources made available to the construction operation carry out the various work tasks. The nature of and the relationships between the work tasks, including the type of equipment and material used by the work force in a construction operation, define the construction technology.

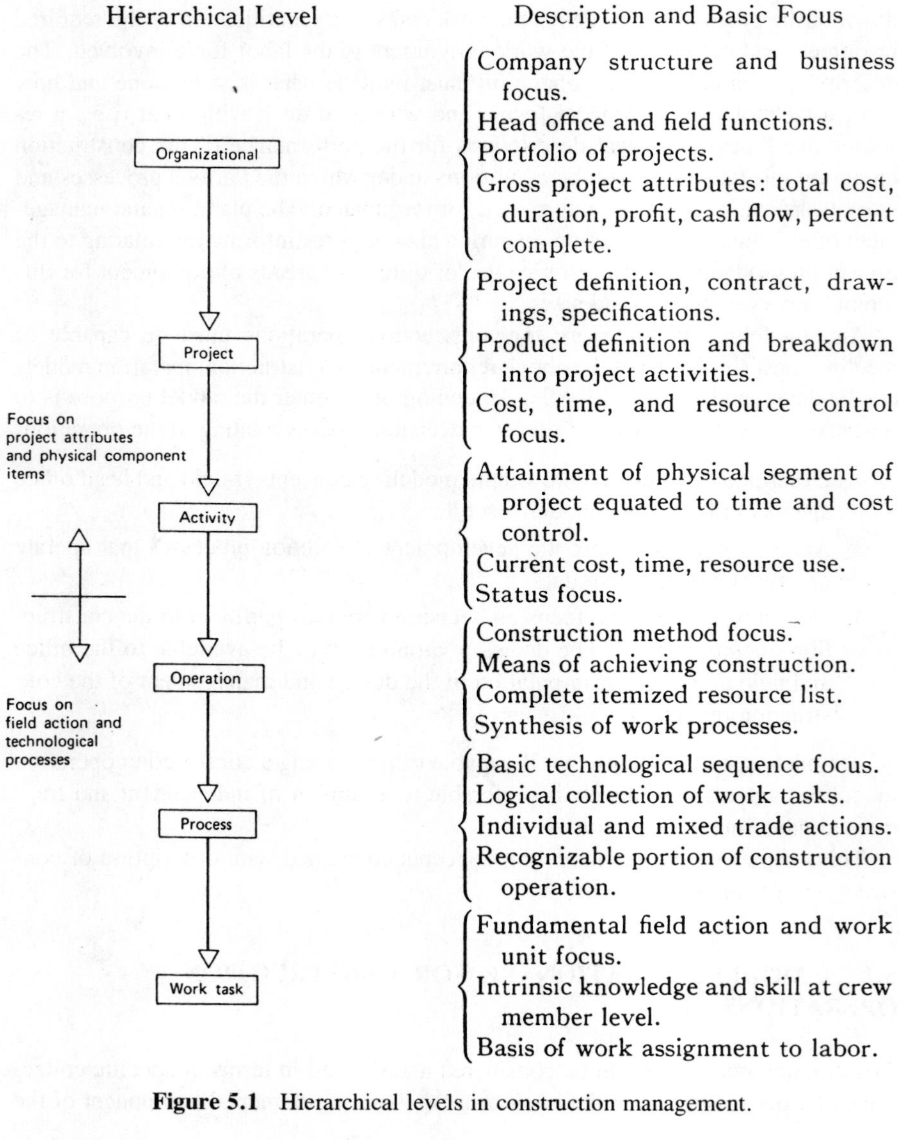

Hierarchical Level   Description and Basic Focus

Organizational
- Company structure and business focus.
- Head office and field functions.
- Portfolio of projects.
- Gross project attributes: total cost, duration, profit, cash flow, percent complete.

Project
- Project definition, contract, drawings, specifications.
- Product definition and breakdown into project activities.
- Cost, time, and resource control focus.

Focus on project attributes and physical component items

Activity
- Attainment of physical segment of project equated to time and cost control.
- Current cost, time, resource use.
- Status focus.

Operation
- Construction method focus.
- Means of achieving construction.
- Complete itemized resource list.
- Synthesis of work processes.

Focus on field action and technological processes

Process
- Basic technological sequence focus.
- Logical collection of work tasks.
- Individual and mixed trade actions.
- Recognizable portion of construction operation.

Work task
- Fundamental field action and work unit focus.
- Intrinsic knowledge and skill at crew member level.
- Basis of work assignment to labor.

Figure 5.1 Hierarchical levels in construction management.

A meaningful description of a construction operation therefore requires the definition of the basic work tasks and the manner in which the available resource entities perform (or sequentially process through) the work tasks. In this sense individual resources can be said to traverse or flow through work tasks. The sequential, relative, and logical relationships between the various work tasks as defined by the various resource flows portray the plan or static structure of the construction operation and technology. The actual working of the operation can then be described by locating and monitoring, from time to time, the various resource entities as they dynamically traverse the static structure of the operation.

At the construction level, the available resources are labor, equipment, material, and the skills shared by labor and management in their knowledge of the operation and the construction technology involved. Construction operations at the work face level are carried out purposely by explicit work assignment allocations to particular tradesmen, crews, and equipment. Furthermore, the orderly progression of the work and the technology of the construction operation will dictate the sequence in which a specific resource (i.e., equipment piece or tradesperson) will carry out each assigned work task.

An initial step in the development of a construction operation model requires the enumeration of the various resource units involved in the operation. Once a resource unit has been defined, it is easy, for a person with a knowledge of the construction operation, to identify the specific work tasks in the operation with which the resource entity is involved, and thereby to determine the sequential ordering of the work tasks.

At any one time, resource units can be considered to be either in an active state (i.e., working at, or traversing, a work state) or to be in idle state (i.e., idle awaiting the opportunity to become active in a succeeding work task). In this way flow units can be considered as passing from state to state whenever the transfer becomes possible. The need for distinguishing between active and passive states is caused by the requirement for the modeling and monitoring of idle and unproductive resources in construction operations as measures of the productivity, or lack of productivity, of the operation.

The identification of the individual resource units associated with a construction operation, the elemental work tasks, and the resource unit flow routes through the work tasks can be made the basic rationale for the modeling of construction operations. The requirement for modeling the conditions that must be met for a work task or process to become active can be satisfied in terms of conditional logic relating passive resource states with active work tasks and the use of special symbols distinguishing between unconstrained and constrained work tasks.

5.2 BASIC MODELING ELEMENTS

A graphical modeling format for work states and entity flow modeling can be developed using three basic modeling shapes:

1. The active state square node representing a work task
2. The idle state circle representing a delay or waiting position for a resource entity
3. The directional flow arc representing the path of a resource entity as it moves between idle and active states

The symbols used (see Fig. 5.2) for each modeling element are designed to be simple and helpful in developing schematic representations of the construction operation being modeled. Two basic shapes (squares and circles) are used to model active and passive resource states; together with directed arrows (arcs) for resource flow direction, they help to provide a quick visual grasp of the structure of a construction operation. These symbols are the basic modeling elements of the CYCLONE (CYCLic Operations NEtwork) modeling system. They are used to build networks of active and idle states to represent cyclic construction processes.

It is convenient to distinguish between the unconstrained (i.e., normal) work task and the constrained (i.e., requiring the initial satisfaction of conditions) work task. While all work tasks are modeled schematically as square nodes, the constrained work task is modeled as a square node with a corner slash. Thus a total of four symbols is required for the modeling of the structure and resource entity flow of construction operations (see Fig. 5.2).

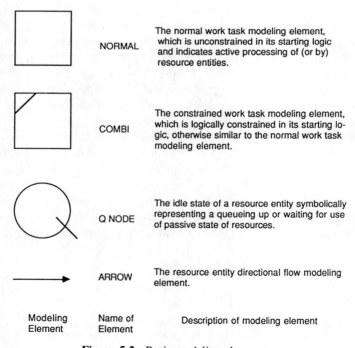

Modeling Element	Name of Element	Description of modeling element
□	NORMAL	The normal work task modeling element, which is unconstrained in its starting logic and indicates active processing of (or by) resource entities.
◿□	COMBI	The constrained work task modeling element, which is logically constrained in its starting logic, otherwise similar to the normal work task modeling element.
Q	Q NODE	The idle state of a resource entity symbolically representing a queueing up or waiting for use of passive state of resources.
→	ARROW	The resource entity directional flow modeling element.

Figure 5.2 Basic modeling elements.

The active working-state models are the NORMAL and COMBI modeling elements. Both have a square-node format and model work tasks. Since the work task is the basic component of a construction operation, it should be chosen so that its name or description is sufficient to convey to a crew member or supervisor the nature, technology, work content, and resources needed to fulfill the work task.

Simple examples of work task activities are breaking open brick pallets, preparing column formwork, and loading trucks with front-end loaders. The definition of a work task thus implies a verbal description of the work task, an indication of the resource entities involved, and an explicit (or implied) awareness of the time commitment of the resource entities to the work task.

5.3 THE NORMAL ELEMENT

A graphical model of a NORMAL work task should have the following ingredients (see Fig. 5.3):

1. A unique graphical format [i.e., a square node]. The square node model of a work task can thus be imagined to be an abbreviated form of the bar-chart representation of an activity.
2. A descriptive label.
3. A user-defined time delay function that prescribes the resource entity transit time through the work task.
4. An indication of the resource entities that must transit the work task.

The square node models the "NORMAL" work task. A NORMAL work task has no ingredience constraints on its resource entities. For NORMAL work tasks, the arrival of a resource entity at the input side is sufficient to allow the commencement of the work task; consequently, there is no queueing up of resource entities before a NORMAL work task. Furthermore, since the NORMAL work task is not ingredience-constrained, several resource entities can traverse it simultaneously. For instance, in modeling a brick plant, a normal work task model can

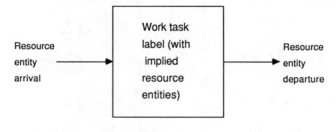

Work task time delay function

Figure 5.3 The NORMAL work task modeling element.

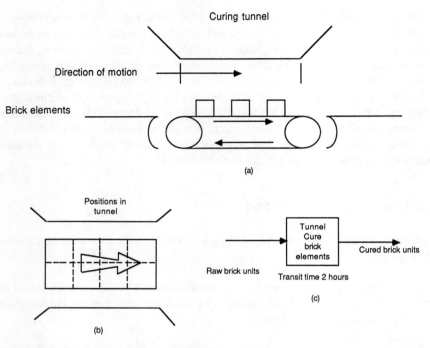

(a)

(b)

Raw brick units

Tunnel Cure brick elements

Cured brick units

Transit time 2 hours

(c)

Figure 5.4 Work task model of precast element tunnel curing: (*a*) elevation view of tunnel; (*b*) plan view of tunnel; (*c*) work task model.

be used to represent the time spent by a cart of bricks in passing through the curing tunnel (see Fig. 5.4). This type of model may be used because several carts can be passing through the work task simultaneously (i.e., in parallel), and the carts are not constrained by a set of requirements or an ingredience condition.

The more general model of a NORMAL work task shown in Figure 5.5 indicates that the work task can be processed by either resource entity A or resource entity B. That is, the work task can commence (or be realized) by the arrival of a resource entity along any of the arcs incident on its input side; thus there is no

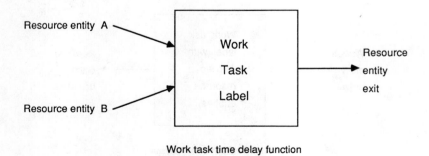

Work task time delay function

Figure 5.5 The NORMAL work task modeling element.

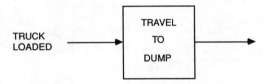

Figure 5.6 A typical use of the NORMAL modeling element.

requirement to wait until ingredience constraints are met. Once the NORMAL work task has commenced, the activating resource is captured for the duration of the user-defined delay. In this way the NORMAL element models the time involvement of resources in work tasks.

A simple example of the use of the NORMAL element is the travel of a truck from the loading area to the dump area in an earth hauling operation. Once loading is completed, the truck can commence travel to the dump site immediately. The only prerequisite of the "Travel to Dump" task beginning is the completion of the preceding work task, "Load Truck." This is an unconstrained sequence not requiring additional resources before travel can begin. Therefore, completion of the preceding element is sufficient to allow commencement of the following element (e.g., Travel to Dump). This is illustrated in Figure 5.6.

5.4 THE COMBI ELEMENT

The COMBI modeling element is similar to the NORMAL modeling element with the additional logical requirement that a prescribed set of input requirements on its input side must be met before the work task can be initialized. Thus, all the logical requirements needed for the COMBI work task must be available for the work task to proceed. An essential modeling feature of the COMBI element is, therefore, the identification and definition of the ingredience set of input resources required to start up the work task being modeled by the COMBI element.

A graphical model of a COMBI work task is shown in Figure 5.7. A simple example is the loading of a truck with earth using a front end loader. This work task requires a truck, a front-end loader, and earth (see Fig. 5.8). These resources

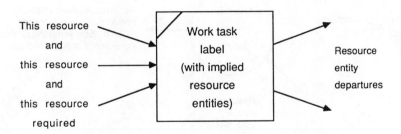

Figure 5.7 The COMBI work task modeling element.

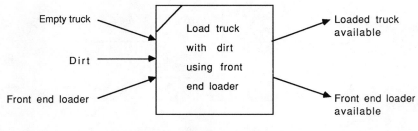

Figure 5.8 The use of the COMBI modeling element.

define the ingredience constraint or the logical conditions required before the work task element "Load Truck with Dirt Using Front-end Loader" can begin.

5.5 THE QUEUE NODE

The idle or passive state of a resource is modeled by a circle with a slash (Q) to denote a QUEUE node (or Q node) (see Fig. 5.9). The Q node is the graphical form; it has an obvious similarity to the concept of units being idle in a waiting line queue. Usually the time a resource entity is in the idle state is unknown and depends on external conditions in a construction operation relating to various work task states and durations and on the efficient design and management of the operation.

The QUEUE node acts as a storage location for resources entering an idle state and releases the resources one at a time to the following COMBI node whenever the COMBI node ingredience logic is satisfied. Figure 5.10 illustrates the idle state of a front-end loader. QUEUE nodes always precede COMBI nodes, and the set of QUEUE nodes associated with each COMBI node establishes its ingredience logic. Thus the COMBI node of Figure 5.8 is related to three QUEUE nodes, as shown in Figure 5.11.

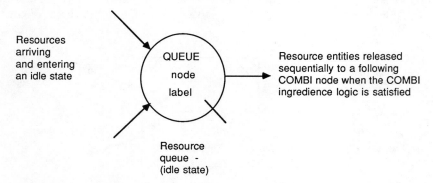

Figure 5.9 The queue (Q node) resource idle state modeling element.

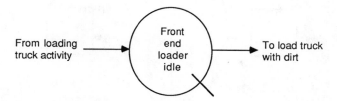

Figure 5.10 The use of the QUEUE (Q node) resource modeling element.

5.6 THE COUNTER

The COUNTER as shown in Figure 5.12 is a special-function node. This element is also referred to as an ACCUMULATOR. In addition to the COUNTER, a multipurpose node or element called the FUNCTION node is available to perform functions such as flow unit consolidation. This will be discussed in Chapter 7. The purpose of the COUNTER is to count the number of times a key unit passes a particular control point in the network model so that production can be measured. A value can also be assigned to the COUNTER element to scale the single arriving unit to reflect the level of production. In an earthmoving example, the multiplier might be the capacity of a truck, say, 15 yd^3. Each time a truck entity passes the COUNTER, the multiplier would increase the cumulative amount of production by 15 yd^3.

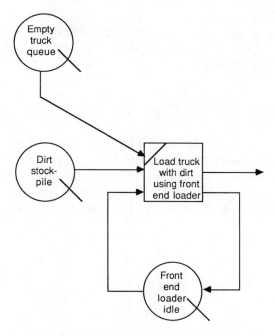

Figure 5.11 The COMBI–QUEUE node ingredience logic.

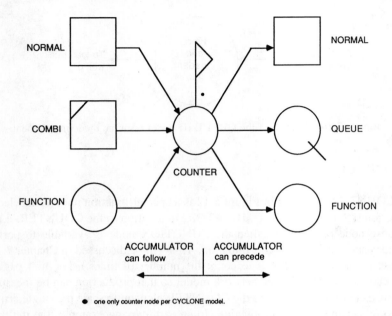

Figure 5.12 The counter modeling element.

The COUNTER element may also be used to control the number of times the system cycles before stopping or shutting down. This means that the duration of an experiment can be controlled by assigning a number of cycles to the counter, which establishes the number of times the COUNTER is to be realized. Once the user established number of units have reached the COUNTER element, the simulation experiment is terminated.

5.7 THE ARROW OR DIRECTED ARC

An arrow or directed ARC (see Fig. 5.13) is used to model the direction of resource entity flow between the various active-state "square nodes" and the passive-state "queue nodes." The ARC modeling element enables the logic and structure of the operation to be developed. The ARC has no time delay properties and simply models entity flow direction. The entity transit time, once initiated, is instantaneous.

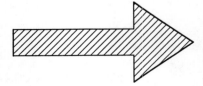

Figure 5.13 The ARC modeling element for resource entity direction flow logic.

The four modeling elements (i.e., the NORMAL, COMBI, QUEUE node, and the ARC) are the basic building elements. These elements can be combined in several ways to model construction operations, as shown in the next chapter.

5.8   DYNAMIC RESOURCE ENTITY FLOWS

The static structure of a construction operation provides the technologic blueprint or plan of how the operation is to be performed. The actual resources assigned to a given operation must traverse through the static structure of the operation. The movement of resource entities through the time-invariant static structure is dynamic and introduces the time-dependent properties of performing construction operations.

During the actual performance of the operation, the current status of the operation and location of the various resource entities in the static structure are functions of time, $f(t)$. Initially, the resource entities must be made available at the input side of the construction operation to start an activity and, finally, on completion of the operation some resources are freed and become available at the output side of the construction operation for future allocation according to the construction plan. During processing, the actual routes and flow patterns that individual resource entities traverse are defined by the static structure. Consequently, the time-dependent resource movement is governed purely by the work task processing durations and the need for resources to wait before initializing ingredience-constrained work tasks. Dynamic resource entity flows are therefore determined by the user-prescribed work task durations and system-generated delays.

The NORMAL and the COMBI have user-defined time-delay functions that prescribe the time period during which resource entities are delayed while processing through these work tasks. The QUEUE node has the potential for storing in a waiting, idle, or queue format the resource entities held up by system requirements pending the satisfaction of COMBI work task ingredience or initializing logic. The ARC modeling element, however, has no time parameters affecting resource entity flow, since it serves purely as a means of indicating resource directional flow logic between the various system NORMAL, COMBI, and QUEUE nodes.

The actual dynamic flow of resource entities through the static structure and their behavior in relation to the modeling elements can be illustrated by considering the resource entities traversing the circuitry or network model of the construction operation. Thus, referring to Figure 5.14a, an entity "1" arrives at the input side of a NORMAL along ARC B and immediately initiates processing of the work task. The work task duration is modeled by the probability function shown in Figure 5.14b, which has a mean duration of 30 min, a duration range between 20 and 40 min, and a standard deviation of 5 min. Entity "1" is captured in the work task for a duration (say, of 28 min) and activates the output side of the work task, where it emerges as two simultaneous entities "5" to satisfy the emanating ARC C and ARC D. In general, each arriving entity "1" along ARC A or ARC B

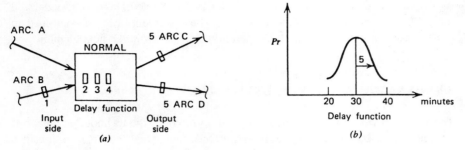

Figure 5.14   Resource entity flows through NORMAL element.

reactivates the NORMAL and requires its own unique duration to process through the NORMAL. Thus entity "2" may require 33-min transit, entity "3" may require 25 min, entity "4" may require 38.5 min, and so forth; consequently, depending on arrival time at the input side of the work task, flow entities may overtake previous arrivals. An example of this situation would occur if the NORMAL element models the transit time of trucks between a dump location and loading point and a faster truck passes a slower truck.

However, if arriving entities cannot pass one another, then Figure 5.14a must be replaced by Figure 5.15a. The entity "1" arrives at the QUEUE node and establishes an entity queue at "2." The entity "2" looks ahead to see if the square-node element that it wishes to process is available; if it is available, it proceeds immediately; if another unit is in the square node (i.e., entity "3"), the entity "2" waits in the QUEUE-node queue format until the square node becomes available (i.e., when entity "3" leaves the work task). Thus, the traditional concept of a queue is established by the QUEUE-node logic, and the Figure 5.15a situation implies that the following work task processor has a special flow characteristic that constrains flow. A more rigorous and informative modeling of the situation is given in Figure 5.15b. Here the square node is replaced by a COMBI node and two ARCS. In this situation entity "2" must queue until the slave entity "4" is avail-

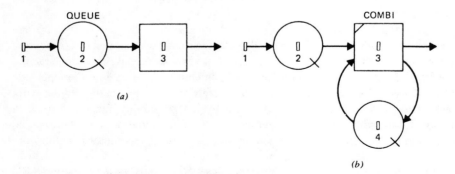

Figure 5.15   (a) No passing and (b) single-transit work tasks.

able in its queue position; otherwise the ingredience logic for the COMBI node (i.e., entity "2" and one entity "4") is not satisfied. Thus Figure 5.15 models the "no-passing" and "single-entity-transit" situation. An example of this situation would be a "Pour Element" work task involving only one form, so that a second element cannot be poured until the first and only form is again available. In this case the slave entity (e.g., 4) would model the form resource unit.

6 Building Process Models

The relative sequence and logic of the work tasks and processes that make up a construction operation constitute the technological structure of the operation. The modeling elements can be used in a variety of patterns to model construction operation structures.

As an example, consider the development of a model for an earthmoving operation that involves the loading of trucks with earth for transport to a dump area. A schematic outline of the operation is shown in Figure 6.1; it uses a front-end loader, some trucks, and earth.

In order to develop the skeletal framework of the earthmoving operation it is necessary to identify the major resources involved (i.e., trucks, front-end loader, and dirt) and establish the various states (i.e., both the active working states and the passive idle states) that the resources traverse in their work assignment paths and cycles. Finally, the integration of the resource paths and cycles establishes the basic structure of the operation.

Each truck, for example, is idle while it queues for loading; it enters active working states when it is being loaded, dumping, traveling loaded to the dump area, and returning empty for another load. A simple model of this work cycle is shown in Figure 6.2a using a single COMBI "Load Truck" work task that requires earth and a front-end loader for initiation; three NORMAL work task elements, "Loaded Truck Travel," "Truck Dump Activity," and "Empty Truck Return"; a single QUEUE element, "Join Truck queue"; and five ARCs indicating the logical relationships between the various truck states.

6.1 MODELING THE STRUCTURE OF CONSTRUCTION OPERATIONS

The front-end loader can be initially modeled by a slave unit cycle involving the active state COMBI element "FEL (front-end loader) Loading," the idle QUEUE element " FEL Idle," and two entity flow directional logic ARCs (see Fig. 6.2b).

In Figure 6.2c, a soil path model is shown that uses a source Soil Stockpile and sink destination Soil Dump QUEUE nodes together with a COMBI work task Loaded into Truck and NORMAL work tasks Transport by Truck and Dumped to portray the soil involvement in active work states. Finally, four directional ARCs are required to develop the path structure.

110

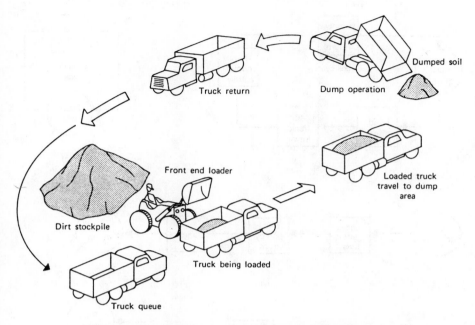

Figure 6.1 Schematic outline of earthmoving operation.

The integrated model incorporating the truck and front-end loader cycles together with the soil path from stockpile to dump is shown in Figure 6.2d. This model can be used as the basis for further development involving dump area spotters and queues, dozer stockpiling operations, and truck maintenance, as well as the basis for further detail such as a fine description of the front-end loader loading cycle.

An extension of the skeletal structure of the earthmoving operation to include dozer stockpiling and spreading operations together with a dump spotter foreman is shown in Figure 6.3. Further development of descriptive models depends on the construction plan to be followed; this development is left to the reader.

The foregoing presentation illustrates that the static (or topologic) structure of construction operations can be developed and illustrated through the proper use and labeling of the basic modeling elements developed in Chapter 5. The static structure associated with the modeling of a construction operation is, in fact, a collection of states (active work tasks and idle queue nodes) together with a set of transformation arcs. The topologic structure, thus defined, is the time-invariant or time-independent structural relationship of the network system model developed for the construction operation. If the work task modeling elements are suitably chosen to be intuitively obvious to the field team and management agents involved in the construction operation, the static structure model is capable of portraying the construction technology and construction method of the construction operation.

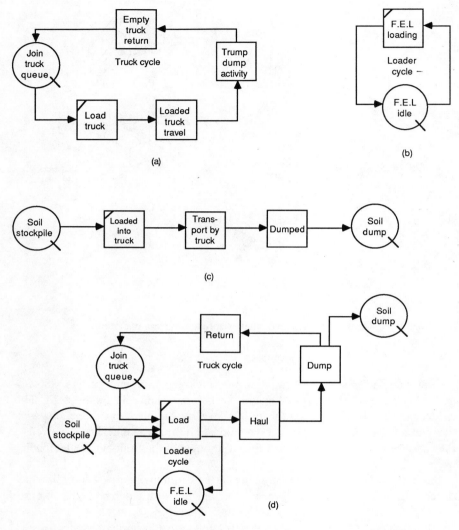

Figure 6.2 Development of operational structure: (*a*) truck cycle; (*b*) loader cycle; (*c*) earthmoving operation.

6.2 THE MODELING PROCEDURE

The procedure for modeling a given construction process involves four basic steps. The steps, as shown in Figure 6.4, are as follows:

1. *Flow Unit Identification.* As a first step, the modeler must identify the system resource flow units that are relevant to system performance and for which transit time information is available or obtainable from the field. The selec-

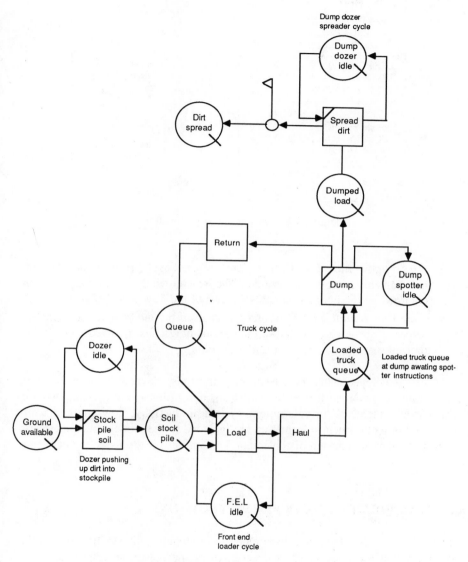

Figure 6.3 Model of earthmoving operation.

tion of the flow entities is very important, since it dictates the degree of modeling detail incorporated into the operation model.

2. *Development of Flow Unit Cycles.* Following identification of the flow units that appear relevant to the process being modeled, the next step in model formulation is to identify the full range of possible states that can be associated with each flow unit and to develop the cycle through which each flow unit passes.

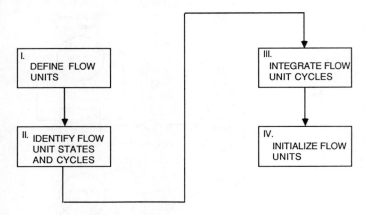

Figure 6.4 Steps in model formulation.

3. *Integration of Flow Unit Cycles.* The flow unit cycles provide the elemental building components of the model. The structure and scope of the model are obtained by the integration and synthesis of the flow unit cycles.
4. *Flow Unit Initialization.* In order to analyze the model and determine the response of the system model, the various flow units involved must be initialized, both in number and initial location.

Models developed using these basic steps must also be modified to provide for monitoring of system performance. This leads to a fifth stage of system design in which special elements for determining system productivity, flow unit characteristics, and other pertinent information are included in the model structure. These aspects of the system design of construction operations are discussed in later chapters.

6.3 PROCEDURES FOR DEFINING FLOW UNITS

As mentioned above, the first step in model formulation is the identification of the flow units that are relevant to or descriptive of the operation or process to be modeled. The proper selection of flow units requires an intimate knowledge of construction operations and establishes the basis for model validity.

Normally the units selected are physical items such as production units (i.e., front-end loader, truck) or resources such as materials, money, and space. Other flow units are informational in nature and indicate that certain conditions have been realized, thus allowing the start of an activity. For instance, a flow unit might indicate that a given inspection has been passed and that the inspected unit can be released for further processing. The availability of a certain amount of money can be represented as a logical unit that allows for the release of a certain amount of work.

The more obvious types of flow units that are of interest in construction are (1) machines, (2) labor, (3) materials, (4) space or location, and (5) informational or logical permits (inspection releases, etc.). All systems at the field operation level involve flow units of these types. A basic step in analyzing any construction operation or process and in developing the appropriate approach and level of detail is to determine which flow units are relevant from the preceding categories.

The machine unit category is self-explanatory and represents the construction equipment appropriate to the operation being modeled (e.g., tractor dozers, cranes, and trucks). In almost all material flow process models, some type of equipment unit is required for movement of materials. In some cases, the equipment elements involved in a construction operation are not critical in constraining the flow of the processed material or flow units and, therefore, need not be included in the model. Model detail, for instance, seldom extends to the level of hand tools as a constraining force in construction processes modeled by management. Normally, only larger pieces of equipment are considered in realistic system models.

Labor resources of all types play a very active role in constraining operations and thus affect the flow of materials and processed units through construction system models. These resources may take the form of an individual laborer or an entire crew. The number of tasks to which labor resources are assigned is also important in defining the associated flow path. The laborer may be assigned only one work task, which leads to the simple "slave" pattern noted previously (Figure 6.2b). On the other hand, a single operator maybe associated with several work tasks in a given operation and constitute a shared resource. Such resources that cycle between various work tasks have a "butterfly" (multiloop) pattern associated with them. A laborer supporting brick masons, for instance, cycles between activities such as (1) tending the mortar mixer, (2) moving bricks to scaffold, (3) carrying mortar to scaffold, and (4) elevating scaffold.

Figure 6.5 shows a "butterfly" flow path for this type of situation. Operators of machinery, crews, and support laborers are examples of the labor type of flow units.

Material flow units are, typically, the units that are processed by a construction system and that, after being processed, become integral parts of the final construction product. The definition of material units is closely tied to the type of system being modeled. If the construction operation under consideration is concerned with the placement of concrete, for example, then the material units may be defined as batch loads or even as cubic yards of concrete. If a road-paving process is being examined the selected productive units may be extended to the level of sections (50 ft, 100 ft) of the road. In any case, these units are discrete quantities and are important in their relation to the productivity measure of the system. The production of the system is measured as some multiple of the units that pass system productivity measuring elements.

Location or space-type flow units usually constrain the access to certain work processes and thus constrain the movement of other unit types. For example, in a brick plant, a large tunnel is used in which the bricks are cured. The bricks are loaded on transit wagons that move slowly through the tunnel, allowing the bricks

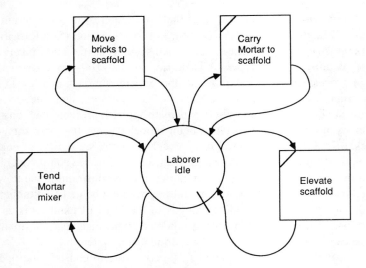

Figure 6.5 "Butterfly" pattern of work task assignment.

to be exposed to curing heat for the prescribed amount of time. The flow rate of the wagons through the tunnel and the size of the tunnel structure govern the number of cart positions or spaces available in the tunnel. A simple model of the brick plant tunnel process is shown in Figure 6.6. In this case, 50 cart positions or space units are associated with the tunnel curing activity and limit the number of brick carts that can be in transit in this activity simultaneously. Therefore, the number of brick carts that can be in activities 2–3–4 is constrained to 50 by the feedback loop 4–1–2. No cart load of bricks can be processed through 2–3–4 until a cart unit is available at QUEUE node 1.

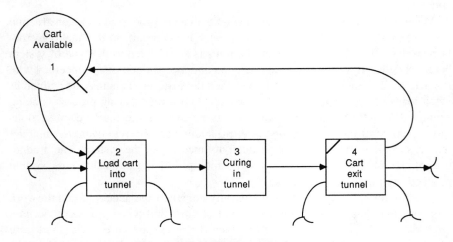

Figure 6.6 Brick plant tunnel process.

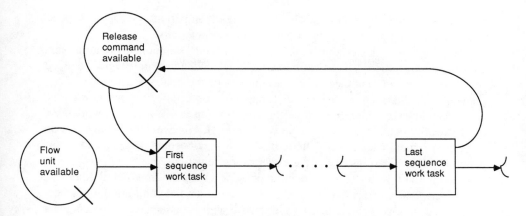

Figure 6.7   Feedback using informational unit.

Similarly, informational flow units are often used to link subsystems and as feedback control mechanisms. In many cases, a unit being processed cannot begin transit of a given set of work tasks until a previous unit has completed the sequence. In this situation, an informational unit can be fed back from the last work task in the set to the initial work task, releasing the sequence and allowing the waiting unit to proceed. An example of this is shown in Figure 6.7.

The logical flow unit is the most versatile of all flow units and, in a sense, encompasses the function of all the other categories. Definition of machine, labor, material, and space units establishes the conditions required for a certain task to begin. The conditions amount to the presence or absence of some resource or release. In some cases, physical items such as money (cash or credit) or documents (purchase orders or invoices) are involved in the realization of the condition. The presence or absence of such units indicates that a condition has or has not been realized and that a release or permit may or may not be granted.

As the first step in construction of a model, the manager must analyze the process to be modeled and determine the flow units from each category described that are relevant and important to the modeling of the situation at hand. Once this is accomplished, the next step in the modeling process concerns the determination of the flow unit cycles and repetitive paths transited by the flow units defined. This begins to establish the cyclic structure of the process network model.

6.4   MODELING FLOW UNIT CYCLES

The identification of all the possible states, both active and passive, that a resource flow unit can occupy establishes the basic framework for modeling the flow unit cycle. The active working states of a resource are usually readily associated with a construction process, whereas the resource idle states are rarely considered. In the initial modeling phase, however, care should be taken to ensure that resource idle states are properly identified. This is particularly important if resource idleness is conditional on the availability of other resource flow units.

The flow unit cycles establish the basic set of model components required in a construction operation. The level of detail with which the flow unit cycle is defined establishes the size, complexity, and usefulness of the final system model. Consequently, at this stage in the model formulation procedure, the purpose of the model should be considered. In some cases, for example, a sequence of work tasks can be readily combined into a single work task. If, however, the behavior of an associated critical resource is thereby hidden, it may become necessary to incorporate the finer detail instead of the grosser representation.

In the initial flow unit cycle modeling phase it will be assumed, until otherwise established, that all work tasks associated with a given flow unit are ingredience-constrained; that is, assume that an active state with its associated time delay is a COMBI element. In this way attention is automatically focused (later during flow unit cycle integration) on the checking of ingredience logic associated with COMBI elements. This results in confirmation of the COMBI node or its replacement and simplification by a NORMAL element.

Since all active states are assumed to be COMBI elements, each one must be preceded by a QUEUE node. This means that for every active state defined in a unit flow cycle, there will be an associated preceding QUEUE node in which the flow unit could be potentially delayed. Following integration of individual cycles into a composite model, a check of each active state is made to determine if any active state is preceded by only one QUEUE node. If this is the case, the modeler reviews the active state defined to establish whether the flow unit is delayed by the requirement for another flow unit before commencing transit of the work task. If another unit is required, this is included by defining its cycle and integrating it into the composite model. If no such constraining unit is required, the work task is redefined as a NORMAL work task, and its preceding QUEUE node is removed (unless the single QUEUE node must be retained to provide a special function such as generation of units—see Chapter 7, Section 7.2).

The flow unit cycles are established by defining the sequence of states through which each of the resources pass. For instance, if the work tasks through which a batch of concrete transits are being considered, the sequence of work tasks might be as follows:

1. Place concrete from bucket.
2. Vibrate placed concrete.
3. Screed placed material.
4. Finish concrete.
5. Cure concrete.

Sequentially ordering the work tasks establishes that the concrete units pass through the "vibrate placed concrete" work task before transiting the screed and finish work tasks. All of these work tasks are considered resource-constrained until otherwise established. In some cases, the cyclic movement of a flow unit through its active states will be easily established. For instance, the cyclic movement of a truck on the haul through the load, travel, dump, and return sequence is not too difficult to identify. In other cases, the system is made cyclic by taking units that are exiting the process (e.g., concrete batches or precast panels) and recycling them in order to reenter them into the system. Having units exit without returning is the intuitive feeling regarding their flow. For instance, if the construction operation is that of placing concrete on the upper floors of a building, the operation involves the concrete arriving at the site in a transit truck and being handled through various work tasks until it is placed in a building component. At this point, the concrete essentially exits the system. In modeling construction operations, however, certain benefits accrue by recycling "used" units. This operation of closing the cycle and the advantages derived will become clearer after a few examples have been examined.

6.5   BASIC RESOURCE ENTITY FLOW PATTERNS

Resource entities move from work task to work task during the performance of a construction operation. The trace or sequence of work tasks traversed by the resource entity together with its idle states indicates the extent of the resource entity's involvement in the construction operation and forms a flow pattern.

A number of basic resource entity flow patterns can be identified (see Fig. 6.8). These basic resource entity flow patterns occur frequently in the modeling of construction operations and correspond to unique recognizable sequences in practice. The patterns can be readily modeled using the modeling elements discussed in Chapter 5.

6.6   THE SLAVE ENTITY PATTERN

Figure 6.8a models a situation that occurs frequently in practice. The slave entity pattern is produced whenever a resource entity is tied to a single active work state in a cyclic sequence. The resource entity endlessly cycles between the active work state and the idle state. In many cases the model is directly valid, while in other cases the resource entity may perform many finer-detail work tasks that are all lumped together into the one work task for the purpose of the construction operation analysis. Table 6.1 offers several practical situations that may be modeled by the slave entity pattern.

6.7   THE BUTTERFLY PATTERN

In many construction operations a resource entity is shared between two (see Fig. 6.8b) or more (see Fig. 6.8c) work tasks. In both situations, once the resource

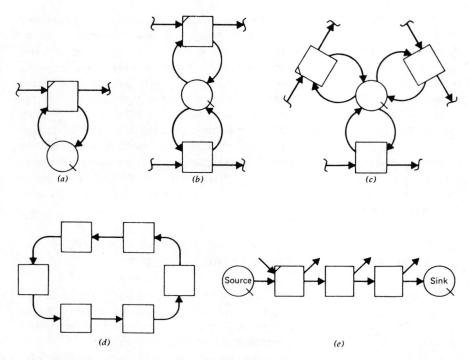

Figure 6.8 Basic resource entity flow patterns: (*a*) slave entity; (*b*) butterfly; (*c*) multiloop work assignment; (*d*) cyclic sequence of work tasks; (*e*) noncyclic or path sequence of work tasks and states.

entity is in the idle state, its subsequent active work task state may depend on any of a number of factors. The availability of the work task and the priority system adopted for the construction operation that gives preference to one work task over another normally control allocation of the resources. Some operational logic may exist for the construction operation that controls the availability of a work task and thus the specific path the resource entity will travel.

6.8 CYCLIC SEQUENCES OF WORK TASKS

In many construction operations repetitive processes exist wherein a resource entity continually transits through a cyclic sequence of states. The cyclic sequence of states may be all active work tasks (see Fig. 6.8*d*) or a mixed sequence of active work tasks and idle states. In Figure 6.8*d* the resource entity processes each work task in turn without delay; in general, the resource entity may be unable to process a work task for any of a number of reasons as explained above and thus be forced into an idle QUEUE node situation.

TABLE 6.1 Slave Entity Patterns

Resource Type	Operation	Active State 1	Idle State 2
Equipment			
Pump	Dewatering	Pump pumping	Pump stationary
Hoist	Hoisting	Hoist in use	Hoist idle
Front end loader	Loading	Loading trucks	Idle awaiting trucks
Hopper	Containing	Hopper full	Hopper empty
Formwork	Forming	Formwork in use	Formwork not in use
Labor			
Mason laborer	Brickwork	Supplying mason with material	Waiting for orders from mason
Laborer 1	Hand excavation by two-man crew	Using pick to loosen material	Waiting for second member of crew to shovel out loosened material
Laborer 2	Hand excavation by two-man crew	Using shovel to remove loosened material	Waiting for first member of crew to loosen material
Material			
Dirt	Earthmoving	Being pushed to heap	Stored in heap
Space			
Working space	Work face	Working space available	Space cluttered up with items and rubbish
Logic			
Permit	Inspection, order, or management	Permission to proceed	Permission unavailable or withdrawn

6.9 NONCYCLIC PATH SEQUENCE OF TASKS

In many construction operations a resource entity is required for a number of work tasks and, when these are completed, the resource entity is no longer needed and exits the operation. Figure 6.8e models this situation where the initial and final QUEUE nodes correspond in concept to source (initial availability) and sink (dispatch) nodes of the resource entity in consideration.

Many examples of the above basic patterns exist in practice. Very often the same operation can be modeled in different ways depending on the level of detail being modeled or the purpose of modeling.

7 System Definition

The CYCLONE modeling formats are designed to provide maximum freedom in assembly logic. The interdependence among elements is reduced to a minimum, and the rules regarding sequential assembly are consequently simple. Table 7.1 summarizes the topologic syntax. The letter in each cell of the tableau indicates the feasibility of an ''A'' element (indicated to the left of each row) preceding a ''B'' element (indicated at the top of each column). The only elements that are logically dependent are the QUEUE node and COMBI processor. A COMBI processor can be preceded only by a QUEUE node and, conversely, the only element that can follow a QUEUE node is a COMBI processor. If the COMBI processor column and QUEUE node row are removed, it can be seen that any precedence relationship among the other elements is acceptable and immaterial with the exception of the ACCUMULATOR element, which cannot precede itself.

7.1 NETWORK STRUCTURE

The actual appearance of the CYCLONE model will depend on the identification and definition of the network elements (i.e., the NORMAL and COMBI) together with the associated QUEUE nodes, ARCs, and logical relationships. However, the network structure depends largely on the way a work task sequence is modeled. Thus, for example, several units can pass through a NORMAL work task in parallel and are not constrained to wait until a processor returns to accompany each unit in transit.

This is very helpful in modeling situations that are not ingredience-constrained. Consider again the curing tunnel situation shown in Figures 7.1*a*, *b*. Carts of bricks are cured using a curing tunnel into which they are loaded and from which they are extracted using a crane. They move from left to right (in the schematic shown) for a predetermined period of time until they are ready for removal. A logical model representing this curing process is shown in Figure 7.1*c*.

The model illustrates the use of a NORMAL work task to provide the transit function representing the brick carts passing through the tunnel. From a systems viewpoint, this is a delay function without ingredience constraint. From a process viewpoint, there is no constraint that prevents the precast element from starting its transit through the tunnel following the loading work task. However, because of the flow pattern of the crane, it is necessary to separate the transit work task from the loading work tasks. The crane must return to its idle state following loading. Therefore, the transit work task cannot be combined with the loading work task

TABLE 7.1   Precedence Tableau of Element A Preceding Element B

A ＼ B	⬠	☐	Q	◯	⚑
⬠	N	I	I	I	I
☐	N	I	I	I	I
Q	M	N	N	N	N
◯	N	I	I	I	I
⚑	N	I	I	I	N

M = required or mandatory
I = immaterial
N = nonfeasible

into a single COMBI element. Insertion of the NORMAL work task allows the brick carts to begin transit immediately following loading, while the crane is returned to its idle state for reallocation. Thus, the network topology is directly affected by the degree of accuracy required in the model.

A similar situation occurred after the Load or Dump work task in the earth-hauling process (see Figs. 6.1 and 6.2). A schematic diagram and a CYCLONE model for a closed loop truck haul situation with a spotter at the dump site is shown in Figures 7.2a and 7.2b. In this case, NORMAL work tasks are used to model the Travel to Dump and Return to Load work tasks. Since transit can commence immediately following both loading and dumping, there is no need to constrain the start of these work tasks. However, following loading and dumping, a resource must exit to an idle state. Therefore it is not possible to combine the load and transit or the dump and transit into a single COMBI element. Figure 7.2c shows an extension of the haul system problem to include probabilistic ARCs that model breakdown situations that can develop on the truck work tasks "Travel to Dump Site" and "Return to Load."

Probabilistic ARCs are used to permit random routing of entities exiting from all nodes other than QUEUE nodes (e.g., COMBI, COUNTER). As an entity exits, a random number is generated that determines the path the entity will take. Referring to Figure 7.2c, entities exiting the "Return to Load" work task break down 3% of the time and are diverted to NORMAL 10. The other 97% of the time, entities exiting from the Dump work task go directly to QUEUE node 8. To implement this action, a random number between 0 and 1.0 is generated. If the random number is less than or equal to 0.03, the exiting unit (truck) is diverted to the Break Down work task. Otherwise, the truck proceeds to the Truck QUEUE at 8. A similar sequence in Figure 7.2c is implemented following the Travel to Dump task (element 3).

Clearly, the addition of probabilistic ARCs builds in alternative sequences of

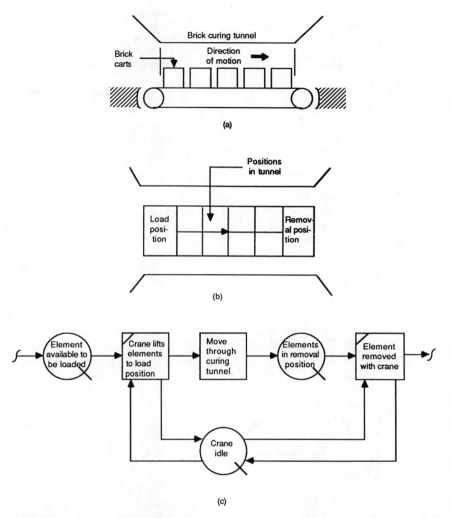

Figure 7.1 Brick plant tunnel: (*a*) elevation view of tunnel; (*b*) plan view of tunnel; (*c*) model of tunnel.

courses of action and thereby affects the topologic structure of the operational model.

7.2 FLOW UNIT INITIALIZATION AND CONTROL

Two elements have associated unit control functions. The QUEUE node defines the points in the system at which flow units can be initialized. Furthermore, units can be generated (as required during the process) at QUEUE nodes by defining a special GENERATE function at the appropriate node. Any idle state location can

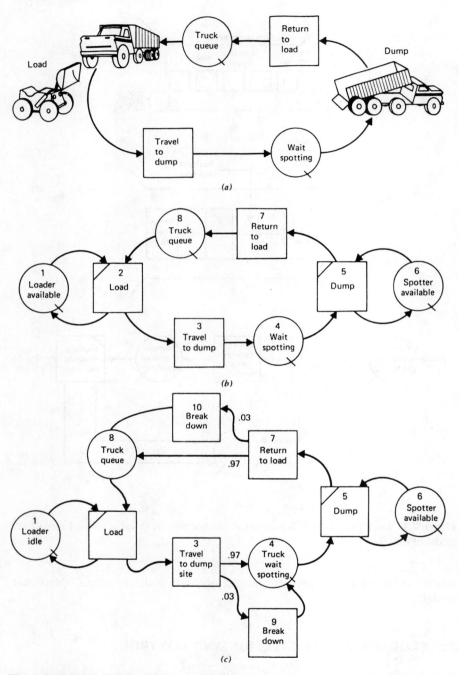

Figure 7.2 Haul operation: (a) schematic diagram; (b) CYCLONE model; (c) haul system with breakdown probabilities.

be used to generate system flow units based on the arrival at the QUEUE node of an incoming unit. It is also possible to aggregate units at a FUNCTION node by defining a CONSOLIDATE function. The location of units in the system at the beginning of a production process establishes the initial conditions of the system. If a loading hopper is full of material at the start of a process, the time required for the system to reach a steady state of operation and the level of production is different from that experienced if the loading hopper is empty. In Figure 7.2, if the trucks are all located at the loading queue when the hauling process starts, the system performance during the early phases of operation will be different from that achieved in which two trucks are located at the "Load" queue and the other two trucks are located at the "Wait Spotting" queue.

Rule: Flow units must be initialized at a waiting position (i.e., QUEUE node) preceding some processor (e.g., trucks waiting at a shovel to load). For this reason, resources as well as processed units are initialized in the CYCLONE system at QUEUE node locations. They must be defined at a QUEUE node preceding a COMBI work task in the network model. The initial position of units flowing in the system is normally indicated in a tabular form that describes unit type (e.g., truck, crew, welder), number of units of each type, and initial QUEUE node location. For clarity, alphabetic characters may be used to indicate each unit type. The letters are shown on the system network model at the initial locations of each unit category or type. Figure 7.3 illustrates the definition of units in graphical format.

Units are initialized in this system as follows:

1. One unit of category A (labor crew) at node 3.
2. Three units of category B (concrete batches) at node 15.
3. Five units of category C (slab forms) at 16.

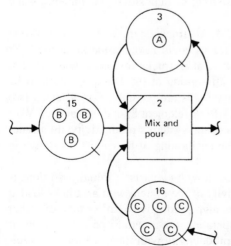

Figure 7.3 Entity graphical representation.

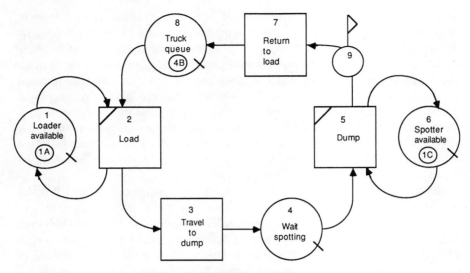

Figure 7.4   Haul unit initialization.

Figure 7.4 gives another example of flow unit initialization in the model introduced in Figure 7.2*a*.

The following units are initialized in the haul system model.

1. One loader unit at QUEUE node 1 (Loader Available).
2. Four truck units at QUEUE node 8 (Truck QUEUE).
3. One spotter (i.e., grade supervisor to "spot" the trucks for dumping) at QUEUE node 6 (Spotter Available).

Again, it should be emphasized that it is illegal to define a unit as starting in an active state. All units must be initialized so as to be in position to commence some work task.

As defined in the system of Figure 7.4, all trucks start in the system in the "Truck" QUEUE waiting to load. The trucks could have started all at the "Wt Spotting" QUEUE node. This would imply that all trucks are loaded with material prior to the start of the system and are all waiting at the Dump location to be spotted. Variations of these two initial conditions are possible. For instance, two trucks could be started in the "Truck" QUEUE (element 8) and two at the "Wait Spotting" (element 4). The initial phases of the system's productivity are a function of the initial conditions defined by the positioning of flow units, as shown in Figure 7.5.

Initial system response for the haul system with two sets of initial conditions is illustrated. It is obvious that the productivity of the system with all trucks loaded and waiting at the Dump location reaches high productivity earlier than the other system. Once the system has been in operation for a certain period of time, it reaches a state in which the variation of productivity with time (d Prod/dt) is rel-

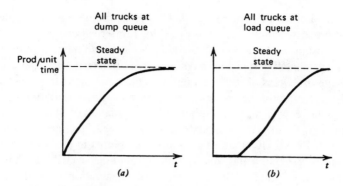

Figure 7.5 Haul system response to different initial conditions: (*a*) four trucks at 4; (*b*) four trucks at 8.

atively small. That is, after a given "transient" period, the system settles down to a steady or nonvariant productivity in which d Prod/dt = O. When the system has reached this time-invariant state, it is said to be in a "steady-state" condition. The time required to reach steady state is a function of the manner in which the units are initialized in the system. This accounts for the variation in productivity levels in the transient period of the haul system.

Care must be taken to insure that all required units in a system are defined. At least one unit must be generated in each flow cycle defined in the system. If a required flow unit is not initialized, the system will at some point call for the required unit, and system execution will be terminated because of failure to meet ingredience requirements. To illustrate, consider again the system of Figure 7.2*a*. If units are initialized, as shown in Figure 7.4, but the spotter unit at 6 is not, all four trucks will process the Load operation and move to 3. Following this, they will all travel to the Wt Spotting idle state at 4, at which point they are unable to proceed because no spotter is available. Then, system execution will stop since ingredience requirements at 5 cannot be met.

7.3 CYCLONE FUNCTIONS

In modeling some productive systems, it is useful to be able to generate or consolidate flow units at certain points in the system. That is, sometimes units representing master flow units break into component units for processing and then the subunits are reconsolidated. A truck, for example, may arrive at an off-loading point where the pallets it is carrying are processed and then shipped further. The truck prior to its arrival at the off-loading point can be considered a single flow unit or master unit. On arrival, it breaks into 20 units that represent the pallets it is carrying and that are to be processed. After processing, the subunit pallets are reloaded onto 10-pallet-capacity trucks and transported further. In order to model this situation a method is needed for generating 20 units to be processed for each arriving truck. Following processing, there is a requirement to aggregate or consolidate the processed units in groups of 10 for further shipment.

In certain instances, an action is initiated after a certain number of cycles of the system or a system subcomponent have occurred. For instance, after a crane has completed five lift cycles of steel framing, it is to be reassigned. A mechanism for counting the cycles and sending a triggering unit to initiate rerouting of the crane is required. This reduces to the counting and consolidation of five cycle pulse units into a single signal unit. To consolidate units, a defined function is available. It is associated with FUNCTION nodes at appropriate points in the network model.

Units can be generated into the system by defining a GENERATE function associated with a selected QUEUE node. Entity generation as well as initiation can take place only at a QUEUE node. Therefore, a GENERATE function can only be associated with a QUEUE node and not with a FUNCTION node. A simplified version of the pallet system is shown in Figure 7.6. In the system just described only one station is available to process the incoming pallets. The model has two processing stations and consists of two QUEUE nodes, a COMBI processor, and a FUNCTION node. Trucks carrying 20-pallet loads arrive at QUEUE node.

A defined GENERATE function is associated with QUEUE node 1; it splits or breaks the arriving truck unit into 20-pallet units (e.g., GEN 20). The pallet processors (two each) cycle between 2 and 3. The use of two processor flow units results in parallel processing. The COMBI processor 3 is ingredience constrained by QUEUE nodes 1 and 2. Following processing, pallet units pass to FUNCTION node 4, where the function CONSOLIDATE operates to group them into 10 parallel loads. After the consolidation of 10 units into one load unit, this single unit moves to QUEUE node 5 and awaits the arrival of a 10-pallet truck at 6.

Graphically, the functions are defined by writing GEN N or CON N under the appropriate QUEUE or FUNCTION node. Therefore, the GENERATE function may be thought of as a discrete entity multiplier and the CONSOLIDATE function as a discrete entity divider. In Figure 7.6, the arriving units are multiplied by 20 and then divided by 10. The amount by which incoming units are multiplied or divided is established by the value of N as defined by the modeler. The ability to multiply and divide units at various points in the system leads to added flexibility

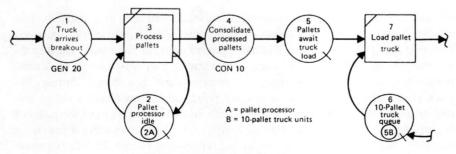

Figure 7.6   Pallet processing model.

and efficiency in modeling, since single units are only expanded as required (e.g., truck to pallets) for processing. Following processing, the expanded units can be reconsolidated to reduce the number of units that must be kept track of at a point in time. This amounts to a means of controlling the population of flow units in the system.

To illustrate this, consider a truck loading problem. On arriving at the loader, the truck is empty and has a 10-unit capacity (e.g., cubic yards or meters). If the loader has a two-unit bucket size, truck capacity is equal to five cycles of the loader. Therefore, on arrival, the truck must generate five load orders to cause the loader to cycle five times. The load orders or commands can also be thought of as five space units representing the empty bay of the truck. Figure 7.7 illustrates schematically (*a*) the arrival of the truck, (*b*) its transformation into five load commands, (*c,d*) the cycling of the loader leading to five loaded spaces, and (*e*) the transformation of these spaces back into a single truck. The five space or command units are present in the system only at the time of processing. At other times, only a single flow unit (the truck) is present. This greatly simplifies the study of the flow units in the system by keeping unneeded detail to a minimum. In simulation of unit flows in the system, great economies can be realized using GEN and CON functions.

Consider another situation in which the CONSOLIDATE function is used alone to trigger an action after a certain number of messages have been received. In many instances, a unit must be rerouted after it has completed a predefined number of cycles. Assume that a crane lifts 10 precast elements into place. After these elements have been placed, a deck-forming sequence can begin. In this case, each time an element is lifted and placed, a count is maintained. After 10 lifts are completed, forming begins. In this case, the CONSOLIDATE function acts to count incoming units and initiate an action when $N = 10$ units (messages) have been received. In the representation of this system component shown in Figure

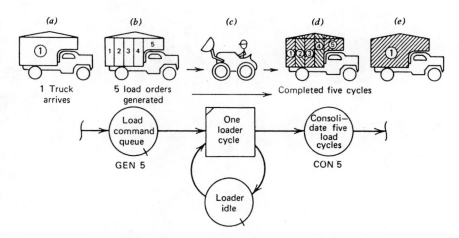

Figure 7.7 Unit control using GEN and CON functions.

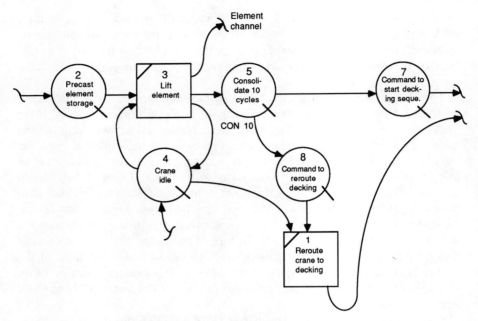

Figure 7.8 Capture–triggering mechanism.

7.8, the FUNCTION node at 5 generates two triggering units. One travels to 7 to signal release of the DECKING sequence and the other travels to 8, initiating the rerouting of the crane to the Decking sequence through COMBI 1, ''Reroute Crane to Decking.'' In this case, QUEUE node 8 initiates the ''capture'' and rerouting of the crane.

Capture–triggering mechanisms of this type have many applications in building construction process models.

8 Extended Modeling Concepts

Previous chapters have introduced the basic CYCLONE modeling elements: the COMBI and NORMAL work tasks, the QUEUE node and the directional arc. Also introduced were probabilistic arcs, GENERATE and CONSOLIDATE functions, and the notion of slave, butterfly, and triggering mechanisms. In this chapter, examples are presented to demonstrate how the basic building concepts are extended to more advanced models. In addition, all the examples in this chapter illustrate some form of divert or capture mechanism.

8.1 MODIFIED EARTHMOVING MODEL

Consider again the earthmoving example of Chapter 6. In that example, a front-end loader was used to load trucks where the soil was stockpiled by a bulldozer. In the current example, the front-end loader is used to load trucks and to stockpile the soil. Parenthetically, since most front-end loaders have a limited cutting capability, it is assumed the soil is a soft to medium-hard material.

An alternative modeling approach is presented in this example. In previous examples, cycles were identified and then integrated into a final network. This alternative approach focuses on work tasks and the resources needed rather than individual cycles. Table 8.1 shows a possible ordering of work tasks and resources for the modified earthmoving example. For simplicity, the spotter and spread dirt cycle of the original model have been omitted.

Beginning CYCLONE modelers are often tempted to include the truck and loader operators as resources. Generally, the number of resources is a function of level of detail of the model. It should be apparent in this instance, however, that the operators are integral to the trucks and loader and should not be considered separately anymore than one would consider fuel or tires as a resource. Moreover, it should be noted that when a resource can be viewed as unlimited (that is, always available), the modeler can sometimes disregard the resource. Using this guidance, the modeler might be inclined to disregard the resource labeled "Material" in Table 8.1. But in the development to follow, it will be shown that the resource "Material" is needed in the logic of the network. The best guidance, therefore, is to keep questionable resources in the network until an analysis justifies dropping them.

Returning to the alternative modeling approach, the next step, after listing the work tasks and resources, is to select a work task from the list and identify the resources needed to begin the work task. For instance, select the work task "Load

TABLE 8.1 Work Tasks and Resources

Work Tasks	Resources
Load trucks	Three trucks
Haul to dump	Front-end loader
Dump	Material
Return	

Trucks.'' The resource entities needed to start this task are a truck, a loader, and material. Additionally, it is appropriate during this step to identify the resources exiting the work task. One exiting entity is clearly the loader. The other exiting entity is a loaded truck. In this instance, the truck entity and a material entity have combined to produce a loaded truck, with the obvious implication the truck is loaded with the material from a QUEUE node ''Material.'' Figure 8.1 shows how the resources and work task are related to develop a segment of the network.

Since the work task ''Load Truck'' requires more than one resource to begin, the work task is necessarily a COMBI and the three resources required to start the work task are placed in QUEUE nodes as shown in Figure 8.2.

Moreover, because ''Loader'' is on both the input and output sides of the ''Load Truck'' work task, it is assumed this resource can be placed in the QUEUE node labeled ''Loader.''

Next, the modeler would repeat the above procedure for another work task. Consider the work task ''Stockpile Material.'' Here, the resource entities needed to start this work task are a loader and material. Entities exiting this work task are the loader and stockpiled material. Figure 8.3 illustrates the entering and exiting resource entities for ''Stockpile Material.''

As before, the work task ''Stockpile Material'' is a COMBI because of the multiple input resources. The resource ''Loader'' appearing on both sides of the COMBI might indicate that it is to be associated with a single QUEUE node. These relationships are indicated in Figure 8.4.

The other work tasks are also evaluated in terms of entering and exiting resource entities. In the case of these remaining three work tasks, some shortcuts may be

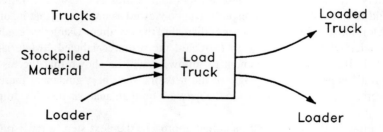

Figure 8.1 Resource entities for loading trucks.

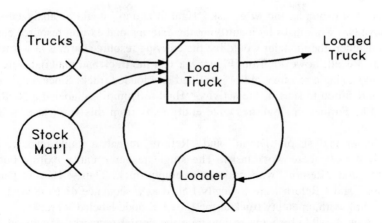

Figure 8.2 Partial network segment for loading trucks.

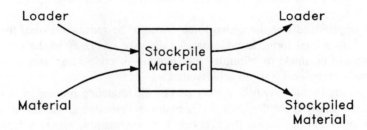

Figure 8.3 Resource entities for stockpiling material.

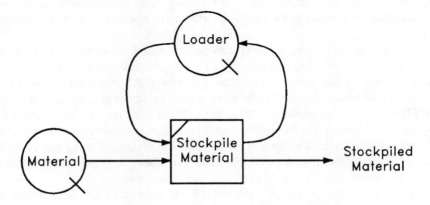

Figure 8.4 Partial network segment for stockpiling material.

possible. For example, the work task "Haul to Dump" could be analyzed just as the above two work tasks by identifying the entering and exiting resource entities. In this manner, the modeler would list the entering resource entities as a truck and material. Recall, however, from Figure 8.1 that the truck and material were combined into a single resource entity identified as a loaded truck. Since only a single entity is required to start it, the activity "Haul to Dump" is labeled a NORMAL work task. Further, a single resource entity exits from this work task: a loaded truck.

A similar analysis of "Dump" and "Return" reveals a single resource entity enters and exits these work tasks. The single resource entity exiting both the "Dump" and "Return" work tasks is an empty truck. Thus, "Haul to Dump," "Dump," and "Return" are logically linked as a sequence of NORMAL work tasks ending with an empty truck at the QUEUE node labeled "Trucks."

It is now possible to begin to integrate the partial networks developed into a final network. An examination of Figures 8.2 and 8.4 suggests that the QUEUE node "Loader" is a common node. Likewise, the QUEUE node "Stock Mat'l" of Figure 8.2 is an abbreviation of the exiting resource "Stockpiled Material" of Figure 8.4 and they can be combined. A trial network is now drawn as shown in Figure 8.5.

It is emphasized that the network may need to be redrawn several times as it evolves into a final form. In addition, for the sake of clarity in the network, an effort should be made to minimize any overlap of connecting arcs and several iterations may be necessary to achieve this.

Integrating partial networks is only part of the modeling process. The modeler should also analyze the logic of the network by visualizing the movement and sequencing of entities within the network. On examination, Figure 8.5 presents a potential problem with respect to the movement of entities. The problem involves the QUEUE node labeled "Material." Generally, all paths in a CYCLONE network need to be closed and the question becomes how best to incorporate this QUEUE node into the network. More specifically, where do resource entities come from so they will pass through the QUEUE node "Material?" Before answering this question, it is useful to follow the sequencing of resource entities through the network.

Note in Figure 8.5 the butterfly pattern associated with the QUEUE node "Loader." It is implicit in this pattern that the loader resource is used by both of the COMBI's "Load Truck" and "Stockpile Material." According to the rules previously discussed for the COMBI work task, if resources are available at every QUEUE node preceding the COMBI, the COMBI can start. In the network of Figure 8.5 an apparently ambiguous condition arises because it may be possible to have resource entities waiting in all the QUEUE nodes preceding the COMBI "Load Trucks" as well as the COMBI "Stockpile Material." Therefore, a choice must be made as to which COMBI to send the loader resource.

This is a situation in which a single QUEUE node (e.g., QUEUE node 3 in this case) is followed by two or more COMBIs. If the ingredience conditions at

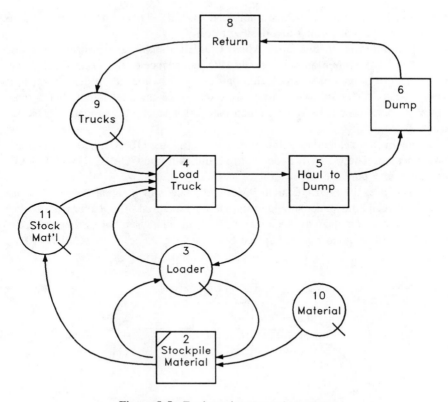

Figure 8.5 Earthmoving network stage 1.

two or more following COMBIs are met, CYCLONE uses a protocol or rule to establish which routing receives priority.

Rule: If a single QUEUE node is followed by two or more COMBINATION elements and the ingredience conditions are met at two or more of the following COMBINATIONS, priority is always given to the COMBINATION with the lowest numeric label (i.e., the lowest numbered following COMBI). Therefore, units in the QUEUE node move to and are combined with other units at the lowest numbered following COMBI node as long as ingredience conditions are met.

In this example, priority will be given to COMBI 2 (Stockpile Material) as long as material units are available at QUEUE node 10. In other words, each time the loader returns to QUEUE node 3, CYCLONE checks whether the ingredience conditions are met at both COMBIs 2 and 4. As long as units are available at QUEUE node 10, based on the rule above, priority must be given to COMBI 2 and the loader will continue to stockpile material. When units are exhausted at QUEUE node 10 (i.e., it is empty), priority will shift to QUEUE node 4.

Whenever COMBIs compete for a common resource entity, CYCLONE sends the common entity to the lowest numbered COMBI. There, the modeler assigns COMBI numbers according to a desired priority.

The question is, in this case, which COMBI should have priority? It would seem that "Stockpile Material" should take precedence over "Load Trucks" because material needs to be stockpiled before it can be loaded on the trucks. Certainly the loader could load a truck as soon as the material is cut, but a more efficient operation results when enough material is stockpiled to load at least one truck.

In any event, a candidate solution on how to link the QUEUE node "Material" has appeared. The QUEUE node could be linked to the COMBI "Load Trucks" as shown in Figure 8.6.

In this case, the QUEUE node "Material" becomes a signal that a truck has been loaded and the loader should stockpile more material. The reader will observe that all the nodes in the network are numbered and a new symbol has been added—a COUNTER at node number 7 (see Section 5.6). At least one COUNTER is

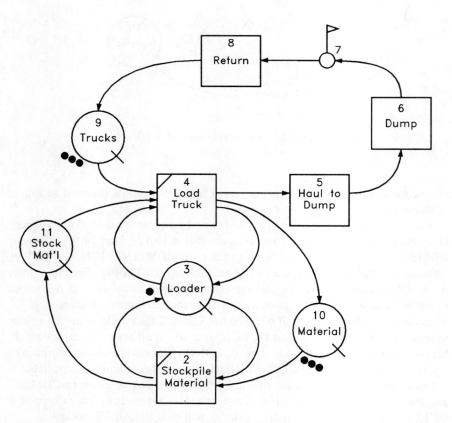

Figure 8.6 Earthmoving network stage 2.

required in any network. As will be seen later, the COUNTER serves two purposes: (1) to gather statistics on the number of resource entities passing through it and (2) to establish a limit on when to stop the computer simulation. Recall from discussion in previous chapters that resource entities flowing through the network do not have to be physical objects. Moreover, it would seem appropriate to rename QUEUE node 10 to "Need for a Load." The final version of the network will reflect this change of name.

Turning again to the analysis of the movement of entities, observe that dots have been placed at three QUEUE nodes representing the configuration of resources at startup: that is, three trucks at QUEUE node 9, one loader at QUEUE node 3, and three entities at QUEUE node 10. Given this placement of resources, the simulation starts at the COMBI "Stockpile Material" because it is the only COMBI whose conditions are satisfied for starting. The "Loader" entity and a "Material" entity move to the COMBI "Stockpile Material." Here, after an appropriate time delay for this work task, one resource entity is released back to the QUEUE node "Loader" and one resource entity is released to the QUEUE node "Stock Mat'l." Now, initial conditions are satisfied for two COMBIs—"Load Trucks" and "Stockpile Material." But, because of the priority numbering, the COMBI "Stockpile Material" starts again eventually releasing entities to QUEUE nodes "Loader" and "Stock Mat'l" as before. This cycle repeats for a third time, finally exhausting resource entities from QUEUE node "Material."

At this point, the three resource entities of QUEUE node 10 have been transferred to QUEUE nodes 11 and activity shifts to the COMBI "Load Truck." But notice, after the first truck is loaded, a resource entity is released to QUEUE node 10. Although it may be preferable to continue loading trucks, the loader is moved to stockpile material because starting conditions of QUEUE node 2 are satisfied and because of the lower number of the QUEUE node. Hence, each time a truck is loaded, the loader is diverted to stockpile material. It might be better, however, to send all three trucks on their way before stockpiling more material. One way to do this is to reverse the priority given to COMBIs 2 and 4 and instead of placing three initial entities at QUEUE node 10, place them at QUEUE node 11. The reader may find it instructive to analyze the sequence of events given these conditions.

Another way to insure all three trucks are loaded before the loader is diverted to cut more stockpile is through the use of CONSOLIDATE and GENERATE functions. Figure 8.7 illustrates where the CONSOLIDATE and GENERATE functions might be added.

Observe the starting conditions for this final version of the network. Only one entity is needed at QUEUE node 10 because the "GEN 3" function now associated with this QUEUE node creates three resource entities for use by COMBI 2. The flow of entities proceeds as before, but the "CON 3" serves to trap one resource entity as each truck is loaded. Only after three trucks are loaded does a single resource entity proceed from the CONSOLIDATE node to QUEUE node 10. On reaching QUEUE node 10, this single resource entity is converted into

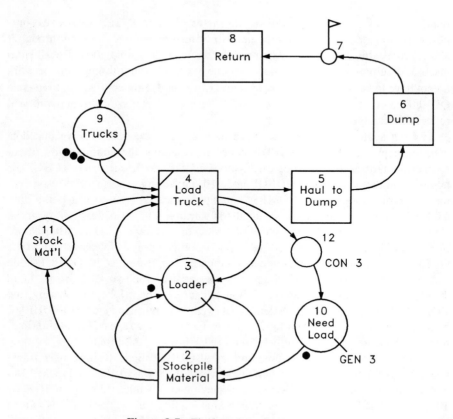

Figure 8.7 Final earthmoving network.

three resource entities. In this way, the model ensures that three truck loads of stockpile are ready before trucks are loaded.

While the preceding discussion illustrates the effect of initial placement of resources, it has not dealt with the time delays of the work tasks. Therefore the question regarding the best mix of resources must take into account the steady-state condition of the model. If the goal of the process is to minimize the waiting time of the trucks, the model must reach steady state before any conclusions can be reached. The steady-state condition for the model in this example depends on the balance of times between the truck cycle and the two loader cycles. Simulation helps provide the answer to the steady-state question by permitting the modeler to analyze the results of various configurations and pick the one best representing the actual physical process.

This example, although relatively simple, has covered several important concepts. One concept is the alternative modeling approach where work tasks are evaluated from the standpoint of resource entities entering and exiting the work task. These individual work tasks are then integrated to form a trial network. The initial position of entities is determined, and the logic of the model is then analyzed in terms of movement of entities.

Other concepts reviewed in the model are the butterfly pattern of the loader resource, priorities numbering for related COMBIs, and CONSOLIDATE and GENERATE functions. It is worth noting at this point that the resource "Material" listed in the resource table was transformed several times during the course of building the model. In one instance, material combined with a truck to become a loaded truck. In another instance, material became the stockpiled material of QUEUE node 11. Finally, material of QUEUE node 10 became a signal that a new load of stockpiled material was needed.

In the following examples, emphasis is placed more on introducing new CYCLONE modeling techniques than on the procedures of model building. The reader might find it instructive to reconstruct the networks below using the model building techniques already presented.

8.2 CONCRETE PUMPING MODEL

This example demonstrates two new concepts: one of a generator of entities into the network; and the other, of a switching mechanism to send entities to an idle COMBI work task. These concepts will be discussed as the example is developed. The process to be considered is a concrete pumping operation. A concrete pump is to be stationed at a job site in preparation for pouring elevated slabs. Transit mix trucks deliver concrete to the pump at specified time intervals. Because of traffic congestion and other delays, it is possible for the transit mix trucks to bunch up on the job site. Assume the configuration of the job site is such that only two trucks can enter the site at one time. These two trucks are positioned on each side of the pump so that when one truck is finished dumping into the pump hopper, the other truck is ready to continue without interruption. The other trucks wait outside the site until a truck leaves, making a space available. As soon as one truck leaves, a truck from outside the site can maneuver to the side of the pump replacing a departing truck.

The network shown in Figure 8.8 models the process described above. Before discussing the switching mechanism of the network, observe the network elements QUEUE nodes 2 and 3 and COMBI 1. This grouping of QUEUE nodes and COMBI depicts a method for inserting entities into the system either on a random basis or at a fixed interval. Entities are inserted on a random basis if the duration of the COMBI is represented by a probability distribution. By the same token, entities are inserted at a fixed interval if the duration of the COMBI is represented by a constant. The time-delay aspects of simulation are discussed further in the next chapter.

Note that both QUEUE nodes 1 and 2 have resource entities waiting as initial

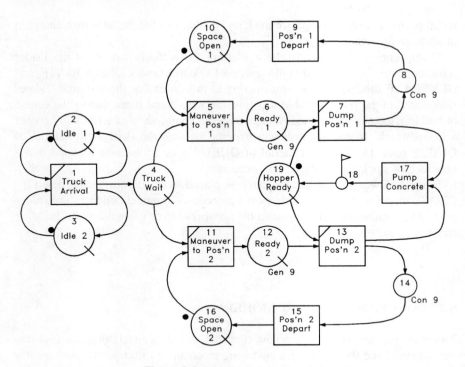

Figure 8.8 Concrete pump network.

conditions. As simulation begins, these two resource entities move to the COMBI "Truck Arrival." After a specified time delay in this COMBI, a transit mix truck is introduced into the system. Also note that both QUEUE nodes 10 and 16 have resource entities as initial conditions. Therefore, the first truck entering the system will move to QUEUE node 5 (the lowest number of two competing COMBIs). Assuming that the second truck is inserted into the system before the first truck leaves, the second truck will move to COMBI 11 because no resource entity is available at QUEUE node 10 (i.e., the space unit at 10 will have been taken by the first truck). At this point, two transit trucks are in position on each side of the pump hopper.

As additional trucks are inserted into the system, they wait at QUEUE node 4 for space to become free at either side of the pump. Following the first truck through the network, it reaches QUEUE node 6, where an appropriate number of hopper loads are generated. As each load is dumped to the hopper, entities are released simultaneously to the pump and the CONSOLIDATE node. When the same number of resource entities has been consolidated as generated, the truck departs, opening a space for another truck beside the pump.

This type of system could be characterized as an assignment of resources on demand system. Demand for the truck resource is created as space becomes free at either QUEUE node 10 or 16. It should be pointed out, however, that this is a special case of assignment of resources on demand. The special case arises because of the symmetric nature of the time duration of the upper and lower cycles of Figure 8.8. In this example, trucks are sequenced alternately between the upper and lower cycles. In other similar (but nonsymmetric) networks, the sequencing of resources might happen randomly according to demand. In contrast, the following example describes a type of systematic or controlled sequencing of resources.

Additionally, the system of Figure 8.8 can be considered to be a nonclosed system. That is, as trucks depart, they leave the system. This represents a situation where, as trucks return to the batch plant, they are not necessarily sent back to the job site. The modeler might choose to design a closed system wherein trucks remain in the network for the duration of the simulation. In this instance, the resource insertion mechanism would be replaced by a concrete batch plant. As trucks depart the job site, they would go to a QUEUE node at the batch plant. The batch plant would then feed truck units back to the pumping network. After the trucks are reloaded, they would travel to the job site and appear once again at QUEUE node 4, where the movement would proceed as described above. In this way, trucks remain in the system continuously. This closed-loop system is shown in Figure 8.9.

8.3   COLUMN POUR MODEL

As noted above, this example illustrates a controlled sequencing of resources rather than a random sequencing. Consider a column pour modeling problem. There are 24 columns to be poured. Each column holds approximately 1.5 yd^3 of concrete. For this example, assume that the columns are formed and only the pouring remains. Prior to each pour, a crew positions movable scaffolding around each column. After each column is poured, the crew moves the scaffold to the next column.

The pouring cycle is as follows. The concrete bucket is filled on the ground and swung into position by a crane. As the bucket approaches the column, the pour crew positions the bucket over the column and releases the concrete. When the concrete is released, the bucket returns to the ground to be refilled and the crew vibrates the concrete just placed. The cycle would then repeat. However, compounding this problem somewhat is the fact that the concrete bucket in this example only holds 1 yd^3 of concrete. This means that after pouring the first cubic yard of concrete, the refilled bucket returns to the same column and only releases 0.5 yd^3. The crew repositions the scaffold to the second column, and the remaining 0.5 yd^3 is released. The bucket returns to be refilled with the 1 yd^3 to complete the pour of the second column. And, the process continues as the crew repositions the scaffold to the third column.

Having defined the process, our next step in constructing this model is to identify any repetitive patterns. A careful examination of the column pour process allows definition of a complete cycle as given in Table 8.2.

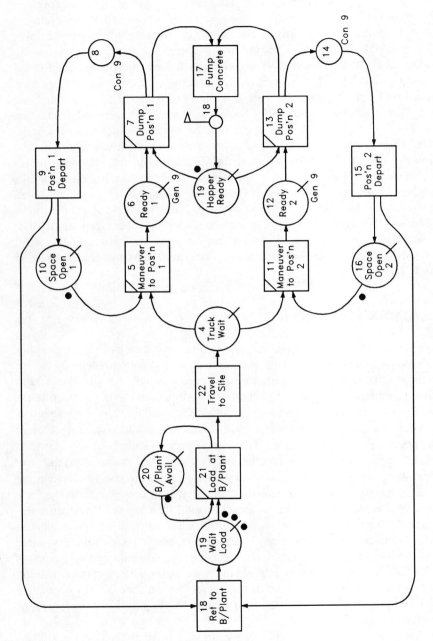

Figure 8.9 Concrete pump network with batch plant.

TABLE 8.2   Column Pour Repetitive Pattern

Work tasks
Load bucket
Pour 1 yd^3
Load bucket
Pour 0.5 yd^3
Reposition
Pour 0.5 yd^3
Load Bucket
Pour 1 yd^3
Load bucket
Reposition

The repetitive pattern is shown graphically in Figure 8.10. Thus the pour of two columns constitutes a complete cycle, and the tasks shown in Table 8.2 repeat after two columns are poured. The modeler must now translate this pattern into a CYCLONE network.

Figure 8.11 shows a CYCLONE network of the column process. Observe that the network does not explicitly model the crane cycle. The crane is assumed to be connected to the bucket throughout the pour operation. In this context, the crane is much like the driver of the vehicle in the example of Section 8.1. The flow through the network is as follows.

As initial conditions, resource entities are available at QUEUE nodes 1 and 2. At startup, these entities move to COMBI 3 where concrete is loaded into the concrete bucket. After loading, the bucket proceeds to QUEUE node 4. Here, there are three possible paths for the exiting resource entity. But, because of the single resource entity at QUEUE node 6, the bucket is directed toward COMBI 5.

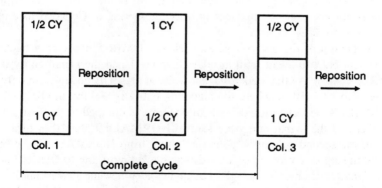

Figure 8.10 Column pour cycles.

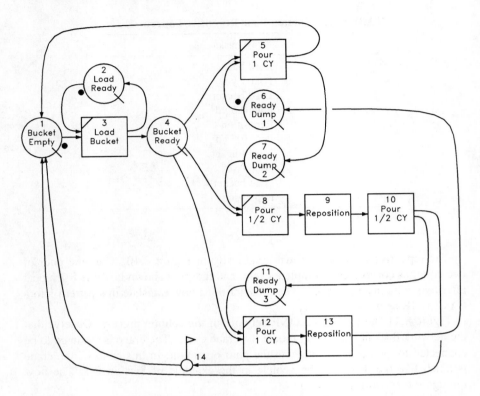

Figure 8.11 Column pour network.

As the discussion proceeds, it will be seen that this single resource entity at QUEUE node 6 is the key to the controlled sequencing of the network and this entity will be termed a signaling entity. Since sequencing will be directed by the signaling entity, it is not necessary to consider priority numbering of COMBIs as a means of controlling the flow.

On completion of the dump concrete task at COMBI 5, the empty bucket resource is sent back to await refill at QUEUE node 1 and the signal entity is sent to QUEUE node 7. As the loaded bucket waits at QUEUE node 4, the only path available is to COMBI 8, where the signaling entity is waiting in QUEUE node 7. At COMBI 8 the remaining 0.5 yd³ of concrete is dumped to the first column. The bucket and the signal entity are sent to NORMALs 9 and 10 in turn where 0.5 yd³ of the second column is poured. Exiting from NORMAL 10, the bucket entity and the signal entity are separated—the bucket returning to QUEUE node 1 and the signal entity moving to QUEUE node 11. When the loaded bucket once more appears at QUEUE node 4, the only path available is to COMBI 12, where the second column is topped off. On completion of COMBI 12, the bucket returns to be filled and the signal entity goes to NORMAL 13 to reposition the scaffolding and then goes to QUEUE node 6. At this point the pattern is ready to be repeated.

The one node not yet mentioned is COUNTER 14. Recall from earlier discussion that the counter serves two purposes: to gather statistics and to stop the computer simulation. Inasmuch as the example called for 24 columns to be poured, a limit of 12 would be assigned to COUNTER 14. This is because each time an entity passes through the counter, two columns have been poured.

The type of control mechanism illustrated in this example might be appropriate to model the installation of drywall, precast panels and similar applications where sequencing of resources is important.

One concluding point is in order regarding this example. Because the purpose of this example is to illustrate controlled sequencing, some detail, such as the truck transit truck cycles, has been omitted for simplicity. Also, as noted earlier, the crane is not explicitly modeled. Likewise, the pour crew and scaffolding are not modeled as resources.

8.4 MATERIAL HOIST MODEL

The focus of this example is on the use of probabilistic arcs. A material hoist services a four-story building as shown in Figure 8.12. Assume that the building shell is complete and the hoist is serving all three upper floors. Admittedly, it would not probably not be economical to use a hoist on a four-story building, but the number of stories is limited in this example to simplify the discussion.

The hoist may have loads for any or all floors, but deliveries are made on the way up the building, not on the way down. The tree diagram of Figure 8.13 illustrates the range of possible choices. For example, the hoist may have a delivery for floor 1 (the first elevated floor). On delivering at floor 1, the next delivery may

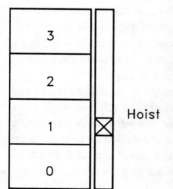

Figure 8.12 Hoist schematic.

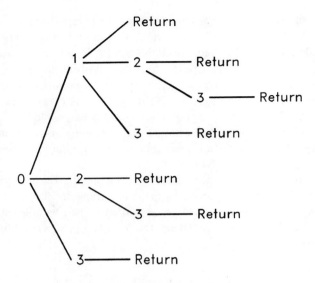

Figure 8.13   Hoist tree diagram.

be for floor 2 or 3 or the hoist may return. Other choices are as shown in Figure 8.12.

The CYCLONE network of Figure 8.14 reflects the logic of the tree diagram of Figure 8.13. Once again, a mechanism for inserting resource entities into the network is used. This mechanism consists of QUEUE nodes 1 and 3 and COMBI 2. The resource entities inserted into the network are in the form of hoist loads. As the first load in inserted into QUEUE node 4, the hoist is waiting at QUEUE node 5. The ingredience conditions for COMBI 6 are met and the loaded hoist transits COMBI 6. In this instance it would be appropriate for COMBI 6 to have a duration of zero since it serves only to route the exiting entity to one of three floors.

The probabilities assigned to each of the arcs 6–7, 6–8, and 6–9 reflect the equal probability of a load needed on each of the three elevated floors. Arc 6–9 is assigned a probability of 0.34 in order for all three probabilities to add up to 1.00. These and other probabilities in this network are assigned in somewhat arbitrary fashion. In an actual application, the modeler would need to analyze the percentages on the basis of observation or possibly from previous experience.

Moreover, in this example, it would seem reasonable to model the travel and return work tasks with constant durations and to model the off-load work tasks with probabilistic distributions. The level of detail of this network is somewhere in the middle of the spectrum. It does not contain great detail to include the hoist operator and labor crews, nor does it lack the detail of travel and off-load times. The next example represents less detail and is included to present an alternative modeling perspective of the hoist problem.

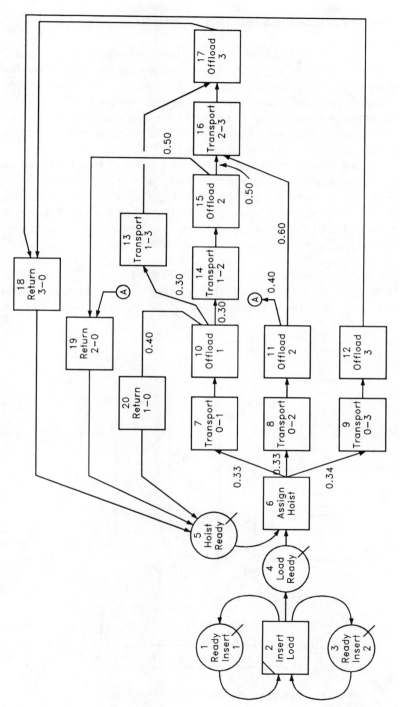

Figure 8.14 CYCLONE diagram of hoist choices.

149

8.5   ALTERNATIVE MATERIAL HOIST MODEL

Consider the same building and hoist system as shown in Figure 8.12, except now the focus is on total time to supply all floors as they are constructed rather than individual floors after the shell is constructed. In this sense, the process can be thought of as nonstationary. That is, the travel time for the hoist will increase as the building progresses—more time is required to service two floors than to service one floor, and so forth. The probability distributions for each phase of construction are shown in a qualitative fashion in Figure 8.15.

Thus, a method must be devised to change the duration of hoist travel as time progresses. In the network of Figure 8.16, this is done with CONSOLIDATE nodes. Initial resource entities are as shown at QUEUE nodes 17 and 20. The process starts at COMBI 19—the only COMBI of the four that could start. The time duration of COMBI 19 is specified by the first probability distribution of Figure 8.15. This distribution represents travel, off-load, and return time for the hoist.

After a specified number of releases, n, from COMBI 19 to CONSOLIDATE node 18, a resource entity moves to QUEUE node 16. However, because COMBI 15 has been assigned a lower number than COMBI 19, the resource entity in

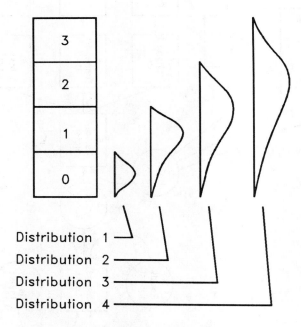

Figure 8.15   Qualitative hoist distributions.

QUEUE node 17 is diverted to COMBI 15 and thereafter sent to QUEUE node 12. COMBI 19 is thereby effectively removed from the network. QUEUE node 20 is directed to COMBI 14 and the second probability distribution becomes operative.

Again, n resource entities are consolidated at node 13. Releases are sent to QUEUE node 11. In a similar way, the process finally moves to the third and fourth probability distributions. Before going to the last example, observe how the numbering system of Figure 8.16 progresses from the end of the network backwards. This is to ensure that divert COMBIs have a lower number than the corresponding hoist COMBIs and will therefore divert, in turn, the hoist entity through to the end of the network.

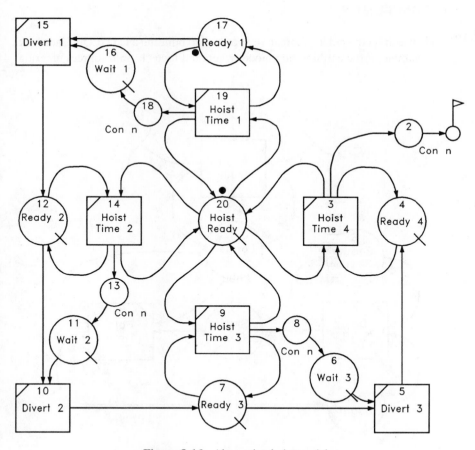

Figure 8.16 Alternative hoist model.

8.6 DAYSHIFT MODEL

This example examines the problem of how to stop the simulation process after a given amount of time and resume operation the next day. Figure 8.17 is a partial network where attention is concentrated on a divert mechanism that stops the simulation after, say, 8 hr. At the beginning of the simulation, the loader in QUEUE node 3 begins to load trucks. At the same time, the resource entities in QUEUE nodes 6 and 7 release to the day shift COMBI 5 where an 8-hr clock starts. At the end of 8 hr a resource entity is sent to QUEUE node 2, diverting the loader from QUEUE node 3 to the hold time COMBI 1. After 16 hr, the loader is released back to QUEUE node 3 and the load truck operation resumes.

8.7 CONCLUSION

This chapter has covered an alternative approach to building a CYCLONE network in the example of the earthmoving model where the focus is on resources entering

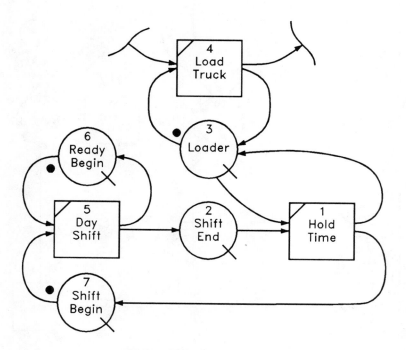

Figure 8.17   Dayshift model.

and exiting work tasks instead of cyclic patterns. The concrete pump model and the column pour model were examples of sequential mechanisms, one not controlled and the other controlled. Also covered in the two hoist models was an approach at modeling a similar process but with different levels of detail. All these models incorporated various divert–capture mechanisms that can be very useful in the modeling process.

9 Modeling Work Task Durations

In CYCLONE models, labor, equipment, and material resources are modeled explicitly as flow units. The technologic characteristics of a construction operation are captured in the structure of the process model, its breakdown into work task sequences, and the flow patterns of units involved in the process. By associating work task transit times with the various work tasks in the CYCLONE model, the model is capable of representing the time period during which resource units are involved with work task sequences. In this way predictive models can be developed for the determination of operational output and productivity over time. It is also possible to determine resource idle times and the influence on productivity of different resource allocations so that management can plan and control the construction operation. Finally, it is possible to determine the workload assigned to specific labor and equipment resources so that meaningful assessments can be made of work quality levels that may be achieved.

This chapter considers how work task durations are determined and used in the CYCLONE system. It investigates the types of time durations that are of interest in modeling construction processes and discusses the manner in which they are defined.

9.1 GENERAL CONCEPTS

Two of the CYCLONE system elements have user-defined delays or transit times associated with them. The COMBI and NORMAL work tasks provide points within the system being modeled at which time durations are defined. These two elements implement the transit time delays input to the system. In this capacity, they function as input elements, allowing the input of system time parameters. None of the other four elements provide for the input of time parameters specified by the modeler. Since the actual system performance is a function of defined element times and system logic, the definition of the COMBI and NORMAL work tasks and their associated time duration is extremely important in capturing the essential features of the real-world system.

Work task transit times determine the time durations that resource flow units are captured by, or are involved in, a work task. These times are determined by estimation or measurement.

Once the operational technology has been decided for a work task, the basic factors that influence work task durations are

1. The magnitude of the work content involved in the work task.
2. The extent to which equipment is used in the work task. Thus the size and efficiency of equipment, equipment characteristics, and functions performed have a dominant effect in equipment heavy operations but less effect in labor-intensive operations.
3. The extent to which labor is used in the work task. Thus the skill level, crew mix, and crew size are significant factors in labor-intensive work tasks. In these cases the intensity of physical effort, team spirit, and motivation directly influence work task productivity and duration.
4. The physical environment of the work site, working conditions, shift hours, weather, and so forth.
5. The level and efficiency of management of the supervisor and at the work site.

The specific influence of these factors on the productivity of a construction operation and thus on the duration of a work task will vary from site to site and is difficult to determine. Some factors are random in their occurrence and impact magnitude, while others have a readily evaluated effect. Depending on the relative magnitude and mix of these factors, work task duration estimates may be deterministic, probabilistic around a known mean, or almost completely random.

Work task duration estimates focus on the determination of a specific duration reflecting the influence of relevant and definable job factors. The remaining unknown or "difficult" factors are bundled together for handling by field management expertise as contingency factors. Thus common practice is to determine specific work task durations with an implicit range of time within which the "normal" duration is tolerated. In this way durations can be considered as either deterministic or probabilistic, depending on the purpose for which the data and the model are used.

For example, in earthmoving operations, the duration of a load, haul, dump, and return cycle for a scraper is affected by haul distance, grades, rolling resistance, engine horsepower efficiency, loading time and operation, altitude, and weather conditions affecting ground surface conditions. Methods exist for determining average (deterministic) cycle times, or probabilistic measures of cycle times (see Sections 9.2 and 9.3).

Depending on the level of detail of the work task definition, the entire earthmoving cycle or a component of it may be considered as a unique work task that requires duration times. In this way deterministic or probabilistic segments may be separated or combined as required for modeling accuracy.

The duration of a labor-intensive work task depends on the energy content required by working conditions and its influence on labor fatigue, the depletion rate demand on human energy reservoirs, the influence of rest periods on energy replenishment, the skill level and planning required for the basic components of the work task, the influence of motivation and attitude to work, team spirit, and the supervisor–crew relationship. Again, the duration of a labor-intensive work task can be determined deterministically through average work rates and productivities or, more realistically, modeled as probabilistic.

In practice work task durations are determined by one or more of the following methods.

1. *Past Experience.* The agents involved in the work process know the time required by frequent previous experience on identical or similar tasks of the same magnitude.

2. *Estimates.* Planning and estimating agents obtain data from previous work on similar work tasks. These data enable them to establish productive rates as a function of the size and mix resources allocated to the work task. Consequently, work task durations can be determined once resource allocation and work content of a work task are established.

3. *Fiat.* In some cases, for large and highly repetitive operations, either past experience is unavailable or unusual features of the planned operation reduce the reliability of past data and it becomes worthwhile to establish experience and data by an initial trial or mock-up run on the operation.

4. *Use of Predictive Models.* Often the basic components of an operation are known and the productivity of resource units working these components is also known. Given situations where the relative magnitude and mix of the basic operation components change, predictive models of productivity and duration estimates are very useful and practical.

The selection of method for the determination of work task duration depends on the nature of the operation and the professional skill of the estimators.

9.2 DETERMINISTIC WORK TASK DURATIONS

A deterministic duration assigns a specific fixed value to the duration of the work task. Any flow unit resource entering the work task is captured for the exact value time period defined before it is released to subsequent system work tasks.

Work task durations may be deterministic for the following reasons:

1. The work task may have a fixed duration or a resource entity is to be captured for a specific time. Simple examples are the mixing time of N revolutions for a concrete mixer, a curing time of one week before stripping, and an 8-hr shift duration.

2. The work task duration may be subject to small variations about a specific mean value so that from any useful time scale the work task duration is constant. Simple examples are hoist time on a building site, scraper cycle times on sites under ideal conditions, and the time to fill a truck with gravel.

3. The purpose of the modeling is such that any probabilistic variation can be ignored. In some cases, as mentioned previously, a broadly defined work task that is subject to large random variation can be broken down into a system set of smaller work tasks, which may localize variations so that some may now be considered as almost deterministic.

Most construction operations performed within a controlled environment, especially those that are equipment-oriented, fall into category 2 above. Most labor-intensive operations fall into category 3 because of individual human factor considerations. Therefore, even in situations where randomness or variability are present, it is common practice to neglect the variation and use a single specific value of time duration when the impact of the variability is considered to be small or insignificant. In such cases, a constant or deterministic value of time duration is selected that adequately represents the time duration of a work task despite any small variation present. Therefore, it is common practice in calculating the productivity of relatively simple construction processes to assume deterministic values for the process work tasks, even though the work tasks may have highly random time durations associated with them.

As mentioned in the previous section, work task durations may be determined in a number of ways. Consider the development of predictive models for equipment-heavy and labor-intensive operation. Simple methodologies have been developed for earthmoving operations based on rated equipment characteristics, equivalent grades, and haulage distances. Similarly, for simple labor work tasks, human factor analysis of microactivities enables models to be developed for the durations of many labor-intensive work tasks.

The simple earthmoving haul operation of the type presented in Chapter 1 is an example of the use of deterministic times. By reducing the model to only two cycles and using deterministic values for the work task durations, it is possible to solve the model simply.

9.3 RANDOM WORK TASK DURATIONS

In systems where the randomness of work task durations and hence of cycle times is considered, system productivity is reduced. The influence of random durations on the movement of resource flow entities causes various entities eventually to bunch together and thus, by arriving at and swamping COMBI-type work tasks, delay the productivity of cycles and operations by increasing the time that resource units spend in idle states pending release to productive work tasks.

In simple cases such as the two-cycle system model of Chapter 1, mathematical techniques based on queueing theory can be used to develop solutions for situations where the random arrival of scrapers to the dozer can be postulated. In order to make the system amenable to mathematical solution, however, it is necessary to make certain assumptions about the characteristics of the system that are not typical of field construction operations. The solution technique that is more general in its application is discrete unit simulation. The concepts employed in discrete unit simulation will be introduced in the next chapter.

Figure 9.1 indicates, in a heuristic manner, the influence of random durations on the scraper fleet production. The curved line of Figure 9.1 slightly below the linear plot of production based on deterministic work task times shows the reduction in production caused by the addition of random variation of cycle activity times. This randomness leads to bunching of the haul units on their cycle. With deterministic work task times, the haul units are assumed to be equidistant in time from one another within their cycle.

In deterministic calculations, all three of the haul units shown in Figure 9.2a are assumed to be exactly 1.35 min apart. In this system, there are three units, and the hauler cycle time is taken as a deterministic value of 4.05 min. In systems that include the effect of random variation of cycle times, "bunching" eventually occurs between the units on the haul cycle. That is, the units do not stay equidistant from one another but are continuously varying the distances between one another. Therefore, as shown in Figure 9.2b, a situation often occurs in which the units on the haul are unequally spaced apart in time from one another. This bunching effect leads to increased idleness and reduced productivity. It is intuitively clear that the three units that are "bunched" as shown in Figure 9.2b will be delayed for a longer period at the scraper queue, since the first unit will arrive to load only 1.05 min instead of 1.35 min in advance of the second unit. The bunching causes units to

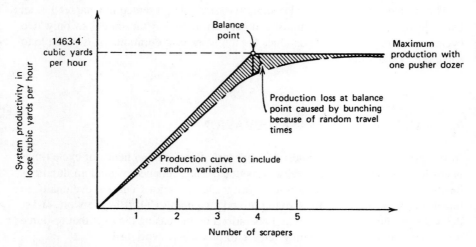

Figure 9.1 Productivity curve to include effect of random cycle times.

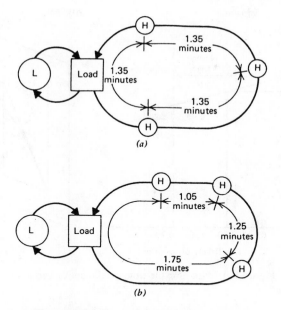

Figure 9.2 Comparison of haul unit cycles.

''get into each other's way.'' The reduction in productivity caused by bunching is shown as the shaded area in Figure 9.1 and is in addition to the reduction in productivity caused by mismatched equipment capacities.

This bunching effect is most detrimental to the production of dual-cycle systems such as the scraper–pusher process at the balance point. Several studies have been conducted to determine the magnitude of the productivity reduction at the balance point because of bunching. Simulation studies conducted by Morgan and Peterson of the research department of the Caterpillar Tractor Company indicate that the impact of random time variation is the standard deviation of the cycle time distribution divided by the average cycle time. Figure 9.3 illustrates this relationship graphically.

As shown in the figure, the loss in deterministic productivity at the balance point is approximately 10% due to the bunching; this results in a system with a cycle coefficient of variation equal to 0.10. The probability distribution used in this analysis was lognormal. Other distributions would yield slightly differing results. The loss in productivity in equipment-heavy operations such as earthmoving is well documented and recognized in the field, mainly because of the capital-intensive nature of the operation and the use of scrapers in both single unit operations and fleet operations. To some extent, field policies have emerged to counteract this effect by occasionally breaking the queue discipline of the scrapers so that they self-load when bunching effects become severe. The resulting increased load and boost time for the scraper adds little to the system productivity, but it does break down the bunching of the scrapers.

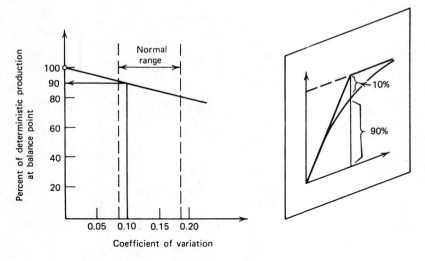

Figure 9.3 Plot of cycle time coefficient of variation.

Many cases exist in construction of the loss of productivity because of the interacting of randomly perturbed cycles. For example, in the masonry operation, initial conditions relating to the status of scaffold stacks of bricks affect mason productivity until a transient ''workup'' phase has elapsed. A common solution to this situation is to have the masons' laborers' workday begin earlier than that of the masons or to have some laborers work later than the masons to restock bricks at the end of a day.

Often the material handling and supply routes on construction sites provide situations where serious interaction develops between apparently totally independent activities. Competition for material hoists and transport space on congested sites reduces productivity, introduces random perturbations in activity durations, and sometimes reaches crisis magnitude. Very little information exists for estimators on the magnitude, influence, and cost of these interactions on construction sites. As will be shown later, CYCLONE models can be developed that focus on idle, queue, or delay aspects of such system problems.

9.4 RANDOM DURATION DISTRIBUTIONS

The impact of random work task durations on multicycle systems is not easily analyzed. It is a function of the numerous configurations that dictate cycle interactions and the types of distribution used to approximate the random variation of activity durations. Probability distributions are useful in describing observed variation of work task times in the field, since they are easily defined in a mathematical sense by relatively few parameters.

Probabilistic functions are used to define the populations from which random time durations can be taken for simulation of a system. A probability density function associates a probability with each of the values of x along the x axis of the function plot. In the case of random time duration, the x variable is the random time delay to be potentially selected. The probability of selection of a particular time duration, x, is given by the area under the curve (i.e., between the curve and the x axis) associated with a specific value.

In the function shown in Figure 9.4, the probability density values are shown along the y axis. The probability that a value will fall at or between 1.0 and 2.0 (i.e., $1.0 < x > 2.0$) is the shaded area. Areas can be approximated by multiplying small segments of the interval between 1.0 and 2.0 by the probability density values that form the height of an inscribed rectangle (see Fig. 9.5).

To get a gross approximation of the area under the curve, simply approximate it as a trapezoid and calculate the area:

$$A = \tfrac{1}{2}b(h_1 + h_2) = \tfrac{1}{2}(1.0)(0.25 + 0.18) = \tfrac{1}{2}(0.43) = 0.215$$

This calculation indicates that the probability that a duration between 1.0 and 2.0 will occur is approximately 21.5%, or a little better than 1 in 5. The probability that a duration of exactly 2.0 occurs is the very small area:

$$A = (0.18)(\Delta x)$$

where Δx is a very small base area defined by the point occupied by 2.0 on the x axis. This base is, of course, infinitesimally small, and the best that can be done is to take a small base and multiply it by the density value as follows:

$$A = 0.18 \times (0.1) = 0.018 = 1.8\%$$

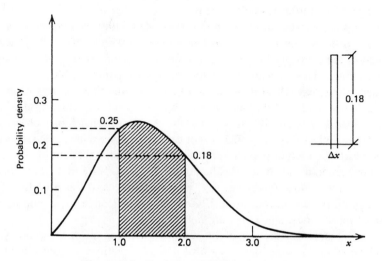

Figure 9.4 Typical probability distribution.

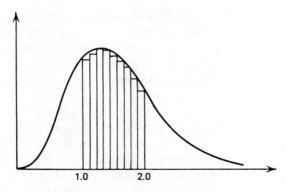

Figure 9.5 Inscribed rectangles.

or, using an even smaller base dimension,

$$A = 0.18 \times (0.01) = 0.0018 = 0.18\%$$

which indicates the probability that the duration will be between 1.95 and 2.05 is approximately 1.8 in 100, or that the chances are around 0.18 in 100 that the duration will be between 1.995 and 2.005.

The rationale for handling the selection of durations in this manner derives from the practice in statistics of organizing observed data in a histogrammatic format. That is, observed data are organized in class intervals. The number of observations in a given interval dictates the height of the histogram segment that has the class interval as base. For example, suppose the times required for a mason to lay a packet of 10 bricks as observed 100 times in the field are summarized in Table 9.1. The observed data have already been arranged into class intervals in the table.

The data plotted as a histogram are presented in Figure 9.6. The y axis of the plot indicates the number of observations associated with each class interval. The number of times an observation occurred between 3.0 and 3.99 is 12. Therefore the height of the element of the histogram with the base 3.00–3.99 has a height of 12 observations. Another way of viewing this is to say that 12/100 is 12% of the 100 observations that fall in the 3.00–3.99 interval, and that the area of this histogram element is 12% of the total area in the histogram. Considering the sample of data collected, it is possible to say that the probability that future observations will fall in this interval is 12%. This is not true in an exact sense, since the sample taken comes from a larger population of values that can be thought of statistically as defining a smooth and continuous function—the probability function of the population of observations consisting, say, of measurement of the time for placing every 10 brick packet placed in the United States. Of course, it is impossible to have all of these times, since they are not recorded.

An approximation of the probability function defining all of these times (i.e., the total population) using our sample of 100 observations can be obtained by drawing a smooth curve through the midpoints of histogram elements, as shown

TABLE 9.1 Field Observations of Bricklaying Times

(1) Interval	(2) Number of Obser- vations	(3) Relative* Frequency	(4) Prob- ability† Density	(5) Cumulative Number of Observations	(6) Cumulative Distribution
0–2.99	0	0	0	0	0
3.0–3.99	12	0.12	0.12	12	0.12
4.0–4.99	15	0.15	0.15	27	0.27
5.0–5.99	18	0.18	0.18	45	0.45
6.0–6.99	21	0.21	0.21	66	0.66
7.0–7.99	16	0.16	0.16	82	0.82
8.0–8.99	12	0.12	0.12	94	0.94
9.0–9.99	6	0.06	0.06	100	1.00
Total	100	1.00			

*Relative frequency = Number of observations in class internal ÷ total operations.

†Probability density = $\dfrac{\text{relative frequency}}{\text{class interval}} = \dfrac{\text{relative frequency}}{1.0}$

in Figure 9.7. In drawing the smooth curve through the element midpoints, assume that the percent area under the curve between 3.00 and 3.99 is still 12% of the total area under the curve. The smooth curve results in loss of the double cross-hatched areas and addition of the singly crosshatched areas. The assumption here is essentially that if the class intervals are small enough, the loss will be balanced by the gain and the area under the curve between 3.00 and 3.99 will still be approximately 12% of the total area under the curve.

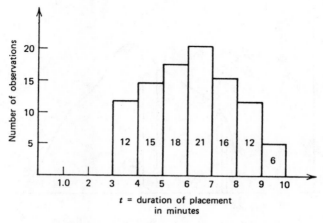

Figure 9.6 Histogram of bricklaying times.

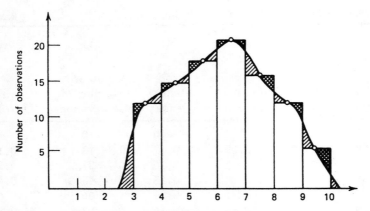

Figure 9.7 Smooth-curve approximation of population probability density (*PD*) function.

The sum of the relative frequencies of the observations in each class interval is 1.0. Therefore the area under the curve representing the probability function of all observations should be 1.0. The value of this area must be set to 1.0; therefore

$$\text{Area} = \int_{x_m}^{x_n} f(x) \, dx = 1.0 \tag{9.1}$$

where x_m = the lower boundary on the *x* axis of the probability curve

x_n = the upper boundary on the *x* axis of the probability curve

$f(x) \, dx$ = area of the element under the curve with base dx and height $f(x)$

This is shown schematically in Figure 9.8.

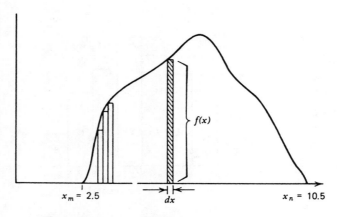

Figure 9.8 Schematic of elements that sum to 1.0.

The actual area between 3.0 and 3.99 that represents the probability that an observation will fall in this interval is given as

$$\text{Area } (x_{3.0-3.99}) = \int_{3.0}^{3.99} f(x) \, dx = P_{3.0-3.99} \tag{9.2}$$

From this definition of probability, the value of the probability density function (*PD*) is

$$PD_i = \frac{P_i}{x_{i+1} - x_i} \tag{9.3}$$

where x_i = the left x-axis boundary of interval i

x_{i+1} = the right x-axis boundary of interval i

$P_i = \int_{x_i}^{x_i+1} f(x) \, dx$ = the probability of an observation in interval i

This is the class interval of 3.00–3.99, which reduces to

$$PD_{3.0-3.99} = \frac{\int_{3.0}^{3.99} f(x) \, dx}{1.0} \tag{9.4}$$

Using the rather large class interval of 1.0 that has been assumed, it is doubtful that $PD_{3.0-3.99}$ or the other values for the smooth curve of Figure 9.7 will equal the histogram probability density values given in Table 9.1. However, if the approach is accepted that the probability density values of the actual population of observations will vary somewhat from the sample, and that the smooth curve is a good approximation for the population, the probability densities given by equation (9.3) are acceptable.

In the generation of random variates for simulation purposes, the distribution that is of greatest importance is the cumulative probability distribution. In the histogram of Figure 9.6, if a plot of the sum of the observations in each element moving from left to right is made, a cumulative probability histogram results as shown on Figure 9.9. The cumulative number of observations is given in column 5 of Table 9.1.

As in the case of the frequency distribution histogram, a continuous function is substituted for this histogram representation. The cumulative probability function is defined as

$$C_r = \sum_{i=1}^{r} p_i \qquad (r = 1, 2, \ldots, n) \tag{9.5}$$

where r = the interval with upper boundary x, to which the probability is to be accumulated ($r \leq n$).

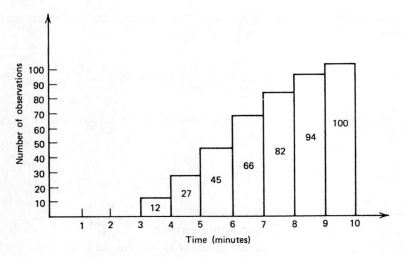

Figure 9.9 Cumulative histogram plot.

Figure 9.7 shows a smooth-curve plot of the probability density figure shown in Table 9.1, column 4. The plot is achieved by associating the probability density values with the midpoints of the class intervals with which they are associated. The cumulative probability associated with interval 3 (5.0–5.99) is then

$$C_3 = \sum_{i=1}^{3} p_i = (0.12 + 0.15 + 0.18) = 0.45$$

The ordinates of this cumulative probability function indicate the probability that an observation (i.e., in this instance, a 10-brick placement time) will be less than or equal to x. Therefore, the ordinate associated with a given x value also represents the area under the probability density curve left of x (since this defines the probability that an observation will occur less than or equal to x). The continuous curve approximating the cumulative probability function for the population of 10-brick placement times is constructed by plotting the cumulative distribution values given in column 6 of Table 9.1 at the x value corresponding to the right-hand (upper boundary) point of the class interval with which they are associated. This is shown in Figure 9.10.

This plot is useful, since it relates all of the probability values from 0 to 1.0 along the y axis to specific values of x (i.e., observation durations). The plot can be used as a nomograph to relate probability values with unique duration values. For instance, if the y axis is entered with the value 0.68, the plot "maps" to a value on the x axis of 7.1 (as shown by the dotted line). Samples can be selected from the distribution plotted in Figure 9.10 by generating values between 0 and 1.0 and entering the cumulative probability distribution on the y axis.

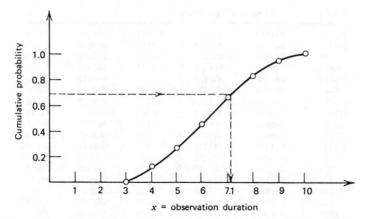

Figure 9.10 Plot of cumulative probability density function.

9.5 MONTE CARLO SIMULATION

The random numbers generated by the Monte Carlo technique act like a ''roll of the dice'' in providing a selection method for generation of random system work task durations. This ''roll of the dice'' concept normally associated with a casino environment leads to the use of the term ''Monte Carlo.'' A major interest in the technique relates to its application in examining and exercising probabilistic or stochastic process models such as those found in construction. In these processes involving random variables, the Monte Carlo process is used to sample the random distributions to generate random time durations or delays. The technique can, however, also be used to evaluate strictly deterministic problems such as the evaluation of complex integrals, which cannot be solved efficiently by analytic mathematical methods.* The validity of response exhibited by systems simulated using Monte Carlo methods depends on the actual randomness of the variates generated. This, in turn, is a function of the randomness of the 0–1.0 range numbers used to enter the cumulative distribution, as shown in Figure 9.10. Random numbers are usually taken from random number lists when performing hand simulations. Such a list of random numbers is shown in Table 9.2. The numbers contained in such tables vary in the number of digits they contain, depending on the source of the table. In this case, the numbers consist of six digits and are scaled to the range 0–1.0, dividing by 1,000,000. The sequence of random numbers required to produce a ''stream'' of random variates is taken from the table by reading across the rows, down the columns, or using any logically nonrepetitive series of numbers.

If a simulation is to be performed requiring a large number of random numbers (e.g., 1000), it is tedious and time-consuming to hand-simulate and utilize random- number tables. In such cases, computer simulation using Monte Carlo methods is utilized for speed and efficiency. It is possible to utilize programs that have

---

*See, for instance, Naylor et al. (1966), ''The Rejection Method,'' Chapter 4.

TABLE 9.2 Random Numbers

258164	244733	824904	959712	284925	062825
547250	466759	943814	751744	707634	376550
279794	797398	656465	505360	241001	256756
676883	778968	934335	028735	444391	538814
056700	668517	599657	172246	663342	229231
339846	006566	593875	032328	975552	373848
036783	039384	559225	193777	846672	240567
220480	236066	351556	161368	074279	441791
321406	414815	106967	967134	445197	647755
926274	486088	641104	796227	668169	882135
551342	913235	842276	771953	004479	286810
304312	473198	047928	626475	026876	718933
823825	835986	287273	754598	161107	308715
937351	010233	721707	522461	965570	850209
617730	061361	325338	131225	786849	095472
702187	367781	949838	786484	715749	572211
208356	204205	692568	713559	289632	429389
248744	223866	150708	276511	735843	573432
490798	341698	903251	657207	410058	436704
941463	047882	413364	938779	457579	617269
642372	286994	477391	626291	742379	699424
849870	720032	861112	753498	449229	191795
093443	315302	160820	515872	692334	149489
560052	889689	963853	091735	149304	895946
356517	332082	776563	549817	894838	369583
136699	990251	654104	295173	362940	215001
819290	934772	920183	769050	175190	288566
910170	602271	514838	609073	049977	729456
454833	609543	085541	650304	299551	371782
725920	653122	512693	897409	795288	228180
350587	914302	072686	378353	766325	367552
101159	479593	435653	267561	592743	202833
606294	874310	610972	603571	552441	215643
633650	239915	661686	617332	310901	292418
797598	437881	965626	699801	863313	752542
780166	624326	787185	194055	174009	510141
675692	741722	717763	163035	042897	057390
049565	445296	301705	977129	257123	343977
297081	668767	808201	856124	541013	061544
780488	008061	843715	130923	242413	368876

large lists of random numbers stored, from which numbers can be called as required. From a computer processing standpoint, however, this is an inefficient approach. It is more common to use a random number generator that generates the latest random number from the previous number. The sequence is derived by "bootstrapping" the new number from the previous one. The first number in the sequence is defined by the user and is called the random-number "seed." The sequence of numbers generated is referred to as a random-number "stream."

Numeric techniques for random- number generation have been improved over the past 50 years. For years, the most widely used method was the *linear congruential scheme* (LCS). An extension of this method, the multiplicative LCS, is a widely used random-number generation technique. This method is defined by the following recursive expression:

$$Z_n = a * Z_{n-1} \text{ MOD } m$$

where Z_0 is defined to be a user selected starting integer (the random- number "seed"), m is the modulus usually defined to be a large integer value (e.g., $2^{31} - 1$), and a is the multiplier and is usually set to the value $7^5 = 16807$.

The random number R_n on the ith iteration is obtained from the expression

$$R_n = \frac{Z_n}{m}$$

To generate random numbers one specifies a seed number Z_0 as a starting value. The value of Z_1 would then be computed resulting in R_1 and so on. It should be noted that the a and m values should be chosen with the utmost care. Otherwise, the random number will start to duplicate after a certain period.

To illustrate the Multiplicative LCS approach, consider the following example. (For this illustration, small values of a and m are used. The values noted previously should be used for practical purposes.) Assume $a = 5$, $m = 7$, and $Z_0 = 9$. The recursive equation will be

$$Z_n = 5 \times Z_{n-1} \text{ MOD } 7$$

$$Z_1 = 5 \times Z_0 \text{ MOD } 7$$

$$= 5 \times 9 \text{ MOD } 7$$

$$= 45 \text{ MOD } 7$$

$$= 3$$

(where 3 is the remainder of the division by 7).

The first generated random number would be $3/7 = 0.4285714$. Continuing this calculational sequence yields the following values:

n	Z_n	R_n
0	9	
1	3	0.4285714
2	1	0.1428571
3	5	0.7142857
4	4	0.5714285
5	6	0.8751429
6	2	0.2857143
7	3	0.4285714

Note that on the seventh iteration the same value of Z was obtained as on the first iteration. This was a result of using a modulus equal to 7. This is why the largest integer value available in the computer typically is used as the random-number generation modulus. The numbers generated are fairly uniform in the range $(0,1)$. If a larger value of m had been used, a denser population could have been gener-

TABLE 9.3 Commonly Used Probability Distributions

Distribution	Formula Defining Probability Density Function	Description	Schematic
Normal	$f(x) = \dfrac{1}{\sigma_x \sqrt{2\pi}} \exp -\dfrac{1}{2}\left(\dfrac{x-\mu_x}{\sigma_x}\right)^2$	The normal distribution is continuous and symmetric about its mean and is defined by two parameters, the mean, μ, and standard deviation, σ.	
Lognormal	$f(y) = \dfrac{1}{\sigma_y \sqrt{2\pi}} \exp\left[\left(-\dfrac{1}{2}\right)\left(\dfrac{y-\mu_y}{\sigma_y}\right)^2\right]$ for $-\infty < y < -\infty$ and $y = \ln x$	The lognormal distribution is continuous and asymmetric and has a mode that is skewed to the mean value. The distribution is characterized by its modal value, its mean, and its standard deviation.	
Exponential	$f(x) = \alpha e^{-\alpha x}$ $\alpha > 0$ and $x \geq 0$	The exponential distribution is continuous in the range $0 < x < +\infty$. It is used to represent the intervals between distinctly random events in "memoryless" processes.	
Gamma (Erlang)	$f(x) = \dfrac{\alpha^k x^{(k-1)} e^{-\alpha x}}{(k-1)!}$ $\left.\begin{array}{l}\alpha > 0\\ k > 0\\ x > 0\end{array}\right\}$ all nonnegative	"If a process consists of "k" successive events and if the total elapsed time of this process can be regarded as the sum of k independent exponential variates each with parameter α, the probability distribution of this sum will be a gamma distribution with parameters α and k."[*] It is a continuous distribution, and may be fitted to many positively skewed distributions of statistical data by varying the values of α and k.	
Poisson	$f(x) = e^{-\lambda}\left(\dfrac{\lambda^x}{x!}\right)$ $x = 0,1,2,\dots$ $\lambda > 0$ or $P(n) = \dfrac{(\lambda t)^n e^{-\lambda t}}{n!}$ λ = mean of exponentially distributed interarrival times	The Poisson is a discrete distribution used to describe the probability of x arrivals in a given time interval, assuming the intervals between arriving units are exponentially distributed with mean $= \lambda$.	

[*]Direct quote Naylor et al; p. 87.

170

ated; that is, the recycling effect would have been delayed and a better spread of numbers throughout the interval would have been obtained.

A truly random number as defined in the mathematical sense is very difficult to generate. Computer-generated random numbers are referred to as *pseudorandom numbers*. Such a number possesses properties that are sufficiently random for use in simulation.

9.6   RANDOM VARIATES FROM CONTINUOUS FUNCTIONS

As mentioned previously, random variates can be generated by entering a cumulative probability function with a uniformly distributed number between 0 and 1.0 and mapping this to the continuous variable axis (i.e., the x axis). This implies that the random variable is continuous in the range of interest and that, therefore, the cumulative probability function is also continuous. It is advantageous to generate random variates from commonly used probability distributions. Use of such standard distributions has the advantage that they can be defined uniquely in terms of relatively few parameters. This does away with the requirement to specify the cumulative function interval by interval, as was done in the masonry problem.

Some of the more commonly used distributions are given in Table 9.3. In general these distributions can be specified using the parameters (1) mean value, (2) standard deviation, and (3) upper and lower limits. A discussion of these parameters and their development from observed data can be found in an introductory statistics text.

10 Simulation

The concept of flow units and their use in the development of network system models for construction operations has been introduced in previous chapters. The identification of work task sequences through flow unit cycles provides the building blocks for the network system model. The network model gives a blueprint (i.e., portrays logic and structure) of a construction operation but gives no indication of the response of the model to the application of specific flow unit resources.

The behavior of a CYCLONE network system model depends on the movement of the resource flow units through the structure and logic of the model. This chapter discusses the mechanics of the flow unit modeling and analysis of CYCLONE models using hand-simulation techniques.

10.1 DISCRETE SYSTEM SIMULATION

Movement of units in the real-world system provides the basis on which the logic of the CYCLONE system network is developed. The object of developing the model of a production system is to examine the interaction between flow units, determine the idleness of productive resources, locate bottlenecks, and estimate production of the system as constituted. In order to achieve this objective the movement of the units through the system must be effected in a manner that simulates the movement of the real-world production resources. This allows the study of the process in an environment that approximates a laboratory.

The use of a model to represent a real-world system is an abstraction and as such loses some of the fidelity and much of the detail of the actual situation. Some very important aspects of the real-world system may be lost in this abstraction. However, care in developing models to an appropriate level of detail should offset this difficulty to a great extent.

The advantages to be achieved focus on the simplicity of a paper and pencil modeling that is inexpensive to construct and may provide insights into system operation that would be extremely expensive to gain by observation of the actual system. These insights often may be gained only after costly mistakes and expensive production runs have been made with the real-world system.

In both cases, the system to be studied must be exercised and observed in order to determine system response and imbalance between resources. Such imbalances result in process bottlenecks and inefficiencies; they should be avoided, or minimized, by a proper selection and balance of resources.

In exercising the model of a construction process that has randomly defined work task duration, Monte Carlo simulation can be used to move the flow units through their cycles and advance them from state to state. As the name implies, the movement of units is achieved by rolling imaginary dice to determine when units are moved and the amounts by which they are delayed. The dice used in this case are the pseudorandom numbers discussed in Chapter 9. These are used to select variates from probability distributions that represent the randomness of system work task durations. If the delays associated with system work tasks are deterministic (i.e., not defined by a probability distribution), the dice are not needed. In such cases, a simple simulation can be performed in which delays at each system work task are predefined.

Whether the durations of work tasks in the construction process to be studied are randomly defined or deterministically defined, the movements of units for purposes of simulation take place at discrete points in time. That is, the simulation work task (e.g., lifting a precast panel with a crane) is defined in terms of its starting event and its end event. During the time between a work task's beginning and end, the units that transit it are captured by the work task and a fix on their location is possible. Therefore, the system network model is concerned only with their movement at fixed points in time (i.e., when they become available for other work tasks). The same is true for those units delayed at QUEUE nodes, where the discrete time at which a flow unit arrives in the QUEUE node and when it leaves are of interest. These points are again discrete events along a line representing the passage of time. Because of the discrete nature of flow unit movements in the systems model, the procedure is referred to as *discrete system simulation*. Procedures for manually moving units in discrete jumps on a graphical model are called *hand simulation*. A discrete computer simulation is implied if a computer is used to move flow units through the system in a discrete manner.

10.2 DISCRETE SIMULATION METHODS

One concept basic to all discrete system simulation is that of a simulation clock (SIM CLOCK), which keeps track of simulation time (SIM TIME). The method by which this clock is advanced is of primary importance, since it establishes how and at what points in SIM TIME the system is reviewed to determine whether flow unit movement should take place. Two methods can be used to advance the SIM CLOCK from discrete point to discrete point in time.

In one method, the clock is advanced in even, equal time steps. In this method, the clock advances from discrete event to discrete event in uniform steps (e.g., 1 min, 2 min, 1 time unit). After each step, the system is reviewed to see if any unit movement was scheduled to take place. This reduces to determining whether any work tasks were scheduled to terminate during the interval of the time step. In other words, a schedule of events is maintained and interrogated across the interval just elapsed to determine whether any units have terminated the work task in which they were delayed. If so, they are released from the work task and allowed to flow further through the system model, generating appropriate delay durations at the following elements to which they pass. This results in new scheduled events that are recorded for future review. This method of advancing the clock is called the *uniform time step* method.

It is an acceptable method when the scheduled movement of flow units is fairly "uniform." However, in cases where there are several quick movements of units followed by long periods of inactivity, this method can be very inefficient. If, for example, a situation occurs in which several units move within 5 min and then no further activity occurs for 2 hr, this method would require 24 reviews of the schedule using time intervals of 5 min (i.e., 2×60 min/5 = 24), during which no system activity would occur. That is, under this method and time interval, the schedule would be consulted 24 times, despite the fact that nothing is scheduled.

A method that is better adapted to the general case is one that advances the SIM CLOCK based on the scanning of end-event times of the work tasks as scheduled during the simulation. In order to implement this method, a simple record-keeping system is used to keep track of scheduled work task end-event times. This system is like an appointment book designed to indicate when work tasks are to terminate. The required system consists of an EVENT list and a CHRONOLOGIC list. When event times are generated, they are listed on the EVENT list. Entries from the event list are recorded in order on the CHRONOLOGIC list at the time they will actually occur.

The EVENT list contains the events that are scheduled, and the CHRONO-LOGIC list contains the events that have occurred. The last event on the CHRO-NOLOGIC list is the event now occurring and therefore represents the time now (TNOW).

The record-keeping aspect of hand simulation keeps track of the event from the time it is generated to the time it has occurred. By examining the EVENT list of Figure 10.1, it can be seen that the event pool consists of four event times that

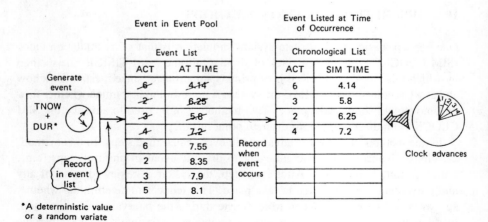

Figure 10.1 Discrete-event processing.

have been generated but that have not yet occurred. Events that have occurred are crossed off the event list and moved to the chronologic list. The SIM CLOCK has not advanced to any of these scheduled events, so they constitute a pool of events that will occur in the future as simulated time progresses. Four events have been transferred to the chronologic list. Three have already occurred, and one (7.2) represents the simulated time now (TNOW). The time steps to date have been 0–4.14, 4.14–5.8, 5.8–6.25, and 6.25–7.2.

Each of the event times generated has been recorded in the sequence generated in the event list of Figure 10.1 and is not listed immediately in the chronologic listing. Some values generated later in the sequence are chronologically earlier than those generated earlier in the sequence. For instance, the third value in the event list is earlier (5.8) than the second value (6.25). If the 6.25 value had been recorded immediately in the chronologic listing, it would have been necessary to erase it and insert the 5.8 value.

The transfer of event times from the event list to the chronologic listing is the mechanism by which the simulation clock is advanced. The last entry on the chronologic listing always indicates the TNOW (time now) value. The event time selected to be moved from the event list to the chronologic list is always the earliest event not yet transferred. Once an event is transferred to the chronologic listing, it is crossed off of the event list. At the moment it is transferred, the previous simulated time is changed to the SIM TIME of the transferred event. In this way it is not possible to get the entries on the chronologic list out of time order from earliest to latest. Since the TNOW value is that of the last event on the chronologic listing, any newly generated event times must be later chronologically, since TNOW + DUR > TNOW. It will be seen later that there are some instances in which the DUR value is zero and the event time generated is immediately transferred through the event list to the chronologic list.

The flow of the simulated event left to right in Figure 10.1 begins at the time the event is generated. This takes place at the time a work task can commence. Durations must be generated for all COMBI and NORMAL elements. The COMBI can commence when its ingredience conditions are satisfied. The NORMAL can commence when units transit from the element or elements preceding it.

One major phase of the simulation of discrete-unit systems consists of the identification of work tasks that can commence, the generation of durations for work tasks that can commence, the calculation of event times corresponding to the termination of these work tasks, and the recording of these termination events in the event pool. This phase is referred to as the EVENT GENERATION phase.

When it is determined that a work task can commence, the first action is to move the flow units that are to transit the work task to the graphical element representing it. After this, the delay time is generated, the work task end or terminal-event time is calculated, and the generated event time is recorded in the EVENT list (i.e., the event is scheduled). At any given TNOW, all work tasks that can commence are started; therefore, they generate END-event times, as shown in Figure 10.2.

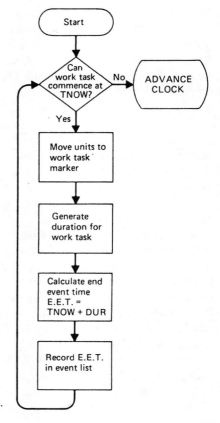

Figure 10.2 Event generation phase.

Once all work tasks that can be started at TNOW are commenced, the second major phase of the simulation procedure is started (i.e., by advancing the SIM CLOCK). This phase is referred to (see Fig. 10.3) as the ADVANCE phase. This phase essentially reduces to the transfer of the next earliest event from the EVENT list to the CHRONOLOGIC list. This makes the next earliest event the last entry in the CHRONOLOGIC list and moves the SIM CLOCK from its previous setting to a TNOW value, which is the SIM time of the transferred event. In other words, the TNOW pointer of Figure 10.1 always drops to the last entry in the CHRO-NOLOGIC list and the SIM CLOCK advances to the event time of the newly inserted entry.

When the SIM CLOCK is advanced, all work tasks that can be terminated are ended and the unit(s) held in transit are released. After all units have been released to the elements following the work tasks in which they have been delayed, the event generation phase recommences. Thus, the continual cycling between the EVENT GENERATION and the CLOCK ADVANCE phases results in the movement of flow units through the simulated system. Since the SIM CLOCK is always advanced to the time of the next earliest schedule event, this procedure for simulation is called the next-event method.

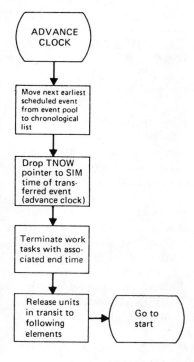

Figure 10.3   Advance phase.

10.3   THE NEXT-EVENT SIMULATION ALGORITHM

The list in Table 10.1 summarizes the steps implied in the flow diagrams shown in Figures 10.2 and 10.3. A composite flow diagram for this procedure is shown in Figure 10.4. A segment has been included in the flow diagram to differentiate between procedures used, depending on whether the delay is deterministic or random. In the case where the work task commenced has an associated random delay, a random variate is generated from the cumulative probability distribution. A flow chart describing this procedure is given in Figure 10.5.

The algorithm described at this point does not contain any provision for developing statistics on performance of the system or the idleness incurred by various units delayed at QUEUE nodes. The algorithm presented provides only for the movement of units through the system defined. Acquisition of statistics will be discussed later in this chapter.

10.4   THE MASONRY MODEL

Consider a mason laborer problem in which the laborer stockpiles the brick pallets on the scaffold. In this system assume that the scaffold is large enough to allow for the stacking of three 10-brick packets. The CYCLONE model for this system

TABLE 10.1 Next-Event Algorithm: A—GENERATION Phase

GENERATION PHASE	1.	CAN ANY WORK TASK START (i.e., are ingredience conditions met)?
	2.	IF YES, continue; if NO, go to step 9.
GENERATE PHASE	3.	MOVE units that can begin transit of a WORK TASK into WORK TASK element.
	4.	CHECK to see if movement of the flow units has caused any clocks to be switched ON, record the unit number (or label), and place a check mark on the ON subcolumn of the CLOCK column of the CHRONOLOGICAL list. If a unit STAT clock has been switched OFF, place a check in the off column, locate the SIM TIME at which the clock was turned on, compute and record the DUR value.
	5.	GENERATE duration of transit time (i.e., transit or processing delay).
	6.	CALCULATE scheduled termination of activity (E.E.T. = TNOW + DUR).
	7.	RECORD E.E.T. in EVENT List.
	8.	Return to step 1.
ADVANCE PHASE	9.	TRANSFER next earliest scheduled event on the EVENT list to the CHRONOLOGICAL list. (Cross out entry on EVENT LIST).
	10.	ADVANCE Sim Clock to time of transferred event.
	11.	TERMINATE work tasks with associated E.E.T.
	12.	RELEASE unit(s) to following elements.
	13.	CHECK STAT clock switches (Same as step 4).
	14.	Return to step 1.

is shown in Figure 10.6. One laborer and three masons are defined in the system to be hand-simulated. The simulation will be carried out until each of the three masons has completed three bricklaying sequences. No bricks are stacked on the scaffold at the beginning of the shift.

Table 10.2 gives the observed time data for resupplying the masons (COMBI 2). Markers representing the laborer, the three stack positions, and the three masons are shown in Figure 10.6 at QUEUE nodes 1, 4, and 7, respectively. Figures 10.7a, b show the cumulative plots of the times for bricklaying and transfer of bricks from the ground stockpile to the stack on the scaffold. Bricklaying times are assumed to be as given in Table 9.1. Assume that the time required by the mason to remove the brick packet (activity 5) has a deterministic value of 1.0 min. Table 10.3 gives the EVENT list and the CHRONOLOGIC list.

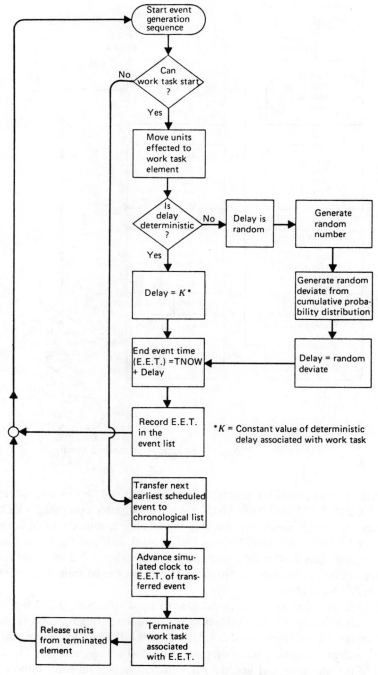

Figure 10.4 Simulation flow diagram.

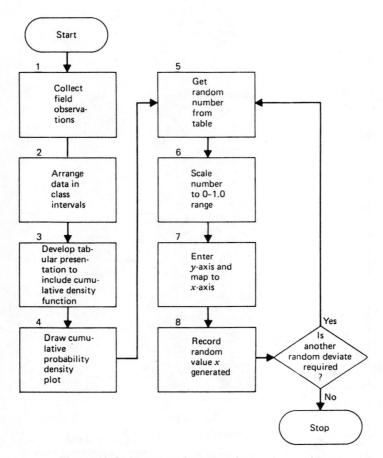

Figure 10.5 Procedure for generating random variates.

In order to understand the simulation process, proceed step by step through the first few GENERATE and ADVANCE cycles. First, after examining which work tasks can commence, notice that the only ingredient requirement that is satisfied is that for COMBI work task 2. Therefore, a stack unit and the laborer unit are moved to work task 2. The time generated for this delay is 3.0 min, and the end-event time (EET) is 3 minutes.* No other work tasks can be started, so move now to the ADVANCE phase.

The EET for work task 2 is transferred to the CHRONOLOGIC list and the clock is advanced to a SIM TIME of 3.0. To indicate this transfer, a check is placed in the TRANSFER column of the EVENT list. At this point, the units delayed in 2 are released, and they pass to the following elements. The laborer returns to the idle state at 1 and the stack marker moves forward to the "POSI-

---

*Since each time a simulation is conducted a new set of random numbers will normally be used, this represents only one of an infinite set of time sequences that could be obtained.

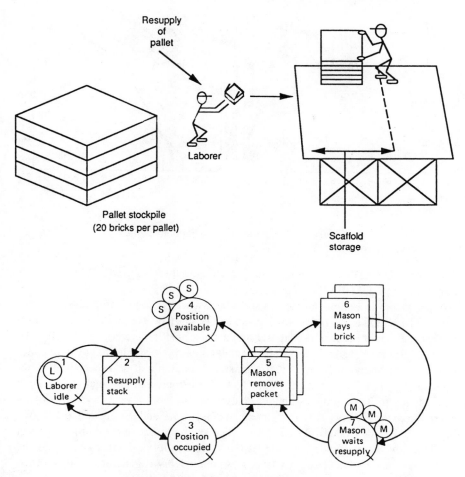

Figure 10.6 Masonry resupply system.

TABLE 10.2 Resupply Times for Mason Problem

Interval	Frequency	Cumula- tive Plots
0.0–0.49	2	2
0.5–0.99	8	10
1.0–1.49	16	26
1.5–1.99	20	46
2.0–2.49	22	68
2.5–2.99	16	84
3.0–3.49	9	93
3.5–3.99	4	97
4.0–4.99	3	100

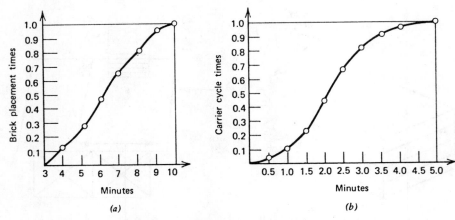

Figure 10.7 Cumulative distributions plots.

TION OCCUPIED'' queue node. Having released these units, simulation control now returns to the GENERATE phase.

The system is again examined to see which work tasks can commence. At this time (SIM CLOCK 3.0), two work tasks can be started. The ingredience requirement for 2 is met because the laborer and a stack position are both available. The "Mason Remove Packet" work task can also begin, since a unit is available at 3 (the unit just released from 2) and a mason is available at 7. Proceeding in accordance with the flowchart of Figure 10.4, the laborer and the stack markers are now moved to 2, generate a random delay time for work task 2, and this time is entered into the EVENT list. The generated delay is 4.9 and that, added to TNOW, gives an EET for COMBI 2 of 7.9 min. The stack marker (representing a packet available on the scaffold) and one of the mason units at 7 are moved into the "Mason Removes Packet" activity at 5. Again a delay is generated and recorded in the EVENT list. The delay in this case is deterministic, since a constant time of 1.0 min was assumed for the mason to pick up the brick packet. Adding this deterministic delay to the TNOW value, an EET for COMBI 5 of 4.0 is obtained. Having started all work tasks that can commence, return to the ADVANCE phase.

TABLE 10.3 EVENT and CHRONOLOGIC Lists (Mason Problem)

Transfer		Event List			Chronological List	
	ACT	TNOW	DUR	E.E.T.	ACT	SIM TIME
X	2	0	3	3	2	3.0
X	2	3.0	4.9	7.9	5	4.0
X	5	3.0	1.0	4.0	2	7.9
	6	4.0	4.5	8.5		

By checking the EVENT list, it can be seen that two events are scheduled to occur. The earlier of these two events is the termination of activity 5 at 4.0 min. Therefore, transfer this event to the CHRONOLOGIC list and place a check for it in the TRANSFER column of the EVENT list. The SIM TIME value of TNOW becomes 4.0, since the clock is advanced to 4.0. Now check to see which work tasks can terminate. The only work task affected by the new TNOW is 5, which is terminated and therefore releases the mason and stack marker flow units. These units pass on to elements 6 and 4, respectively. The movement of the stack marker to 4 indicates that the stack position is now empty. At this point there are no packets stacked on the scaffold. This concludes the ADVANCE phase and requires a return to the GENERATE phase.

Again examining the work tasks that can commence, notice that only one work task can be started. Work task 6 is a NORMAL work task and can start as soon as a unit from 5 arrives. At TNOW (SIM TIME = 4.0), no other work tasks can start. Therefore, a random delay time is generated for NORMAL work task 6. The random delay generated is 4.5. This added to the TNOW value yields an EET for work task 6 of 8.5. This is recorded in the EVENT list.

Returning to the ADVANCE phase, it can be seen that of the two scheduled events that have not been transferred, the earlier of the two is EET = 7.9 associated with work task 2. Transferring this event to the CHRONOLOGIC list, place a check in the TRANSFER column of the EVENT list, and advance the clock to a SIM TIME value of 7.9. That is, TNOW is 7.9. Now terminate work task 2 and allow its flow units to pass forward to the appropriate elements. The stack marker passes to QUEUE node 3, indicating that a packet is now available on the scaffold, and the laborer returns to the idle position at 1 pending redeployment.

If a snapshot of the system and the flow unit positions was made at this point, it would appear as shown in Figure 10.8. No units are being either resupplied (ACT 2) or picked up (ACT 5). The laborer is idle at QUEUE node 1. Two stack positions are empty at QUEUE node 4. That is, a packet is available at QUEUE node, 3, while one mason is placing brick in NORMAL work task 6. This is summarized in Table 10.4.

The discrete time jumps (clock advances) that have taken place to move the system to this configuration are summarized in the TIME LINE diagram shown below the model in Figure 10.8. Three discrete jumps have occurred. In jump one the clock is advanced from 0 to 3.0. Then, the clock moved to 4.0 and finally, it is advanced to the present TNOW value of 7.9. This can also be determined by consulting the SIM TIME column of the CHRONOLOGIC list. This column always shows the sequence of discrete jumps used to move from $t = 0$ to TNOW. The TNOW value is the last entry in the CHRONOLOGIC list. At this point the EVENT list indicates that only one event is scheduled to terminate. This verifies the fact that only one unit is in transit. The only work task containing a transiting flow unit is work task 6, "Mason Lays Brick."

Continuing the simulation through three complete mason cycles results in the EVENT and CHRONOLOGIC lists, as shown in Table 10.5. The system status as of TNOW = 29 is shown in Figure 10.9. By examining the EVENT list, it can

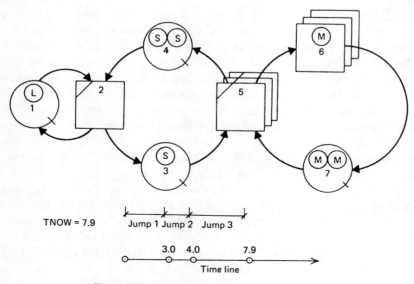

TNOW = 7.9

Figure 10.8 System status at TNOW = 7.9.

be seen that three events are scheduled. Two bricklaying work tasks (6) are in progress. COMBI 2 will end at 32.0. Furthermore, it can be established that another event will be scheduled during the GENERATE phase, since a mason is idle and a packet is available at QUEUE node 3. Therefore, before the clock is advanced from 29.0, a "Pick up" work task (5) EET can be generated and added to the EVENT list.

It is also possible to make a rough estimate of the system productivity to this point. Since three mason cycles have been completed, the number of bricks placed to this point is estimated as follows:

$$3 \text{ masons} \times 3 \text{ cycles} \times 10 \text{ bricks per cycle} = 90 \text{ bricks}$$

TABLE 10.4 Flow Unit Positions at TNOW = 7.9

Element	Flow units at TNOW = 7.9
QUEUE node 1	One laborer unit
COMBI 2	None
QUEUE node 3	One stack unit
QUEUE node 4	Two stack units
COMBI 5	None
NORMAL 6	One mason unit
QUEUE node 7	Two mason units

TABLE 10.5 EVENT and CHRONOLOGIC Lists for Three Mason Cycles

Transfer	Event List					Chronological List		
	ACT	TNOW	DUR	E.E.T.		ACT	SIM TIME	
√	2	0.0	3.0	3.0	(1	2	3.0	
√	2	3.0	4.9	7.9	(2	5	4.0	
√	5	3.0	1.0	4.0	(3	2	7.9	
√	6	4.0	4.5	8.5	(4	6	8.5	1
√	2	7.9	1.0	8.9	(5	2	8.9	
√	5	7.9	1.0	8.9	(6	5	8.9	
√	2	8.9	1.4	10.3	(7	5	9.9	
√	5	8.9	1.0	9.9	(8	2	10.3	
√	6	8.9	4.1	13.0	(9	2	11.3	
√	6	9.9	7.7	17.6	(10	5	11.3	
√	2	10.3	1.0	11.3	(11	2	11.8	
√	5	10.3	1.0	11.3	(12	2	12.3	
√	2	11.3	0.5	11.8	(13	6	13.0	2
√	6	11.3	7.0	18.3	(14	5	14.0	
√	2	11.8	0.5	12.3	(15	2	14.4	
√	5	13.0	1.0	14.0	(16	6	17.6	3
√	2	14.0	0.4	14.4	(17	6	18.3	4
√	6	14.0	8.0	22.0	(18	5	18.6	
√	5	17.6	1.0	18.6	(19	5	19.3	
√	5	18.3	1.0	19.3	(20	2	21.4	
√	2	18.6	2.8	21.4	(21	6	22.0	5
√	6	18.6	3.5	22.1	(22	6	22.1	6
√	6	19.3	3.9	23.2	(23	5	23.0	
√	2	21.4	2.6	24.0	(24	5	23.1	
√	5	22.0	1.0	23.0	(25	6	23.2	7
√	5	22.1	1.0	23.1	(26	2	24.0	
√	6	23.0	6.0	29.0	(27	5	25.0	
√	6	23.1	4.7	27.8	(28	2	25.5	
√	5	24.0	1.0	25.0	(29	2	27.5	
√	2	24.0	1.5	25.5	(30	6	27.8	8
	6	25.0	7.5	32.5	(31	5	28.8	
√	2	25.5	2.0	27.5	(32	6	29.0	9
	2	27.5	4.5	32.0				
√	5	27.8	1.0	28.8				
	6	28.8	6.0	34.8				

This has consumed 29.0 min, and the projected hourly production rate of the system is

$$\frac{60 \text{ min/hr}}{29 \text{ min}} \times 90 \text{ bricks} = 186.2 \text{ bricks/hr}$$

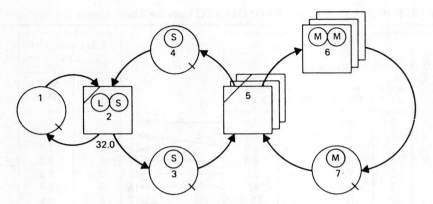

Figure 10.9 System status at TNOW = 29.0.

This estimate of productivity is influenced by the fact that the system is still in a transient stage and has not reached a steady-state level of operation.

An estimate of the length of time that units were waiting in each of the QUEUE nodes can also be determined. By investigating the EVENT list, it can be seen that during the periods 12.3–14.0 and 14.4–18.6, nothing was scheduled at 2. Therefore, the laborer was idle. The fact that work task 2 was busy except for these periods is established by investigating all TNOW and EET values associated with work task 2. These are excerpted and shown in Figure 10.10. There is a continuous pattern of activity at COMBI work task 2 except for the periods noted. This means that during the first 29.0 min of the system simulation, the laborer was idle for 5.9 min, or 20.34% of the time. Similarly, the mason idle time can be established by

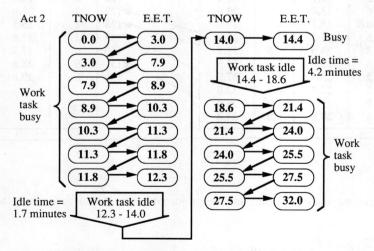

Figure 10.10 Schedule pattern for COMBI work task 2.

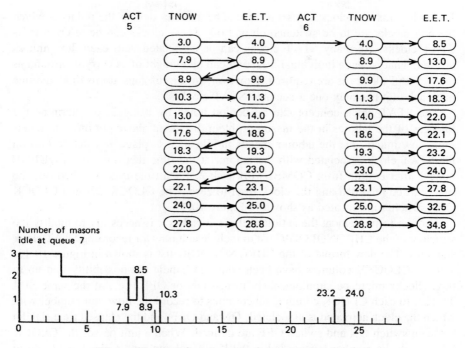

Figure 10.11 Schedule pattern of work tasks 5 and 6.

examining the schedule patterns of activities 5 and 6. By examining Figure 10.11, it can be seen that the only time at which all three masons are idle is during the first 3.0 minutes.

10.5 HAND-SIMULATION STATISTICS

Use of the schedule pattern in determining the idle time for QUEUE nodes is cumbersome. It is also desirable to be able to measure other statistics indicative of the system's overall response and performance. In order to do this it is necessary to expand the hand-simulation algorithm to encompass the concept of statistics collection.

The simulation clock has been discussed in the context of its functions as controller of system time. It is now useful to define additional clocks that are used for measuring system time statistics. These clocks are associated with each flow unit and travel through the system with their assigned unit. They are essentially stopwatches that can be switched on and off to measure flow unit transit time between selected points in the system. The flow unit clocks are designed to be started as their associated flow unit passes one point in the system and then stopped at a later point. The measured time is then recorded and maintained for statistical purposes.

Following this, the clock is reset to zero. The modeler defines the points at which the unit clocks are to be switched on and off. These clocks can be referred to as STAT clocks. As many STAT clocks can be associated with each flow unit as desired. That is, each individual flow unit can be thought of as carrying around as many STAT clocks as are required. However, for discussions at this time, assume that each unit has only one associated STAT clock.

The STAT clock concept can be demonstrated by using it to determine the idleness of the laborer in the mason problem discussed above. In order to determine how much time the laborer spends in idle state 1, place a switch to turn on the STAT clock associated with the laborer at the time that unit enters QUEUE node 1 after exiting from COMBI work task 2. Each time it is switched off, the time will be recorded and the clock reset to zero. The CLOCK ON and CLOCK OFF switches are located as shown in Figure 10.12.

In order to implement the action of these switches it is necessary to modify the structure of the CHRONOLOGIC list to include columns for recording STAT clock statistics. The new format of the CHRONOLOGIC list is shown in Figure 10.13. Three "CLOCK" columns have been added to handle the possibility that up to three clocks might be simultaneously turned ON or OFF (i.e., at the same SIM TIME). In each CLOCK column, subcolumns to record the flow unit number with which the clock activated is associated, ON and OFF, and the elapsed time (DUR) between switch ON and switch OFF are defined. When a unit passes the CLOCK ON switch, its number is recorded in the UNIT column, and a check is placed in the ON column. When the unit passes the OFF switch, a check is placed in the OFF column, and the DUR is calculated and recorded. The DUR is simply the difference between the SIM TIME at which the unit trips the OFF switch minus the SIM TIME at which the ON switch was activated; that is

$$DUR = EET \ (OFF, \ UNIT \ N) - EET \ (ON, \ UNIT \ N)$$

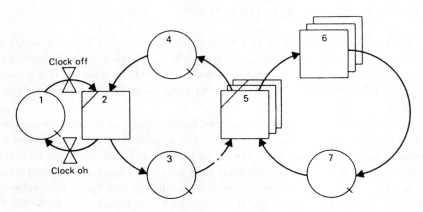

Figure 10.12 STAT CLOCK switches.

A C T	SIM TIME	Chronological List											
		Clock—				Clock—				Clock—			
		Unit number	On	Off	Dur	Unit number	On	Off	Dur	Unit number	On	Off	Dur
2	3.0	L1	√	√	0								
5	4.0												
2	7.9	L1	√	√	0								
6	8.5												

Figure 10.13 CHRONOLOGIC list with STAT CLOCK columns.

The first four lines of the CHRONOLOGIC list of Table 10.5 have been transferred to illustrate the clocking procedure. The completion of work task 2 ($t = 3.0$) releases the laborer flow unit (labeled L1) to pass to QUEUE node 1. Since the ingredience requirements for COMBI 2 are met, the unit transits immediately to 2, tripping the OFF switch. To note this, the OFF column is checked and the DUR value in this case is 0. The same sequence of events occurs at $t = 7.9$, and the L1 unit STAT clock is turned ON and OFF at $t = 7.9$, yielding a zero DUR value.

The CHRONOLOGIC list to include the CLOCK columns for the hand simulation of Table 10.5 is reproduced as Table 10.6. The table shows the ON and OFF switching for the L1 clock throughout the 29.0 min of the hand simulation. Examination of the CLOCK column verifies the same results obtained using the schedule pattern. The STAT clock for L1 is tripped ON and immediately OFF for all transits through QUEUE node 1 with two exceptions. These exceptions at $t = 12.3$ and $t = 14.4$ result in DUR values of 1.7 and 4.2.

10.6 HAND-SIMULATION ALGORITHM WITH STATISTICS

Implementation of the STAT clock concept leads to the addition of some steps to the hand-simulation algorithm presented in Table 10.1. Each time units move in the system, a check must be made to determine whether any unit clocks are switched ON or OFF. This means that it is possible that clocks are tripped ON or OFF at points in the GENERATION phase and the ADVANCE phase. Following

TABLE 10.6   CHRONOLOGIC List with CLOCK Columns

		Chronological List												
		CLOCK____				CLOCK____				CLOCK____				
ACT	SIM TIME	UNIT	ON	OFF	DUR	UNIT	ON	OFF	DUR	UNIT	ON	OFF	DUR	
2	3.0	L1	✓	✓	0									
5	4.0													
2	7.9	L1	✓	✓	0									
6	8.5													
5	8.9													
2	8.9	L1	✓	✓	0									
5	9.9													
2	10.3	L1	✓	✓	0									
2	11.3	L1			0									
5	11.3													
2	11.8	L1	✓	✓	0									
2	12.3	L1	✓											
6	13.0													
5	14.0	L1		✓	1.7									
2	14.4	L1	✓											
6	17.6													
6	18.3													
5	18.6	L1		✓	4.2									
5	19.3													
2	21.4	L1	✓	✓	0									
6	22.0													
6	22.1													
5	23.0													
5	23.1													
6	23.2													
2	24.0	L1	✓	✓	0									
5	25.0													
2	25.5	L1	✓	✓	0									
2	27.5	L1	✓	✓	0									
6	27.8													
5	28.8													
6	29.0													

Total Laborer idleness = 5.9 minutes

step 3 in the procedure in Table 10.1, a check should be made to see if the movement of the flow unit causes the tripping of any STAT clock switches. Similarly, following step 12, at which time units are released to flow to following elements, a check of the unit movement to establish whether STAT clocks are turned ON or OFF should be made.

A flow diagram for the revised procedure is shown as Figure 10.14. The revised algorithm for hand simulation now contains 14 steps and is shown in Table 10.7. In order to understand the new algorithm, consider the first few steps of the original simulation as presented in Section 10.4 to see how its operation implements the functioning of STAT clock switches placed around the idle state 1 of the masonry problem.

TABLE 10.7 Revised Hand Simulation Algorithm

GENERATION PHASE	1. CAN ANY WORK TASK START (i.e., are ingredience conditions met)?
	2. IF YES, continue; if NO, go to step 9.
GENERATE PHASE	3. MOVE units that can begin transit of a WORK TASK into WORK TASK element.
	4. CHECK to see if movement of the flow units has caused any clocks to be switched ON, record the unit number (or label), and place a check mark on the ON subcolumn of the CLOCK column of the CHRONOLOGICAL list. If a unit STAT clock has been switched OFF, place a check in the off column, locate the SIM TIME at which the clock was turned on, compute and record the DUR value.
	5. GENERATE duration of transit time (i.e., transit or processing delay).
	6. CALCULATE scheduled termination of activity (E.E.T. = TNOW + DUR).
	7. RECORD E.E.T. in EVENT List.
	8. Return to step 1.
ADVANCE PHASE	9. TRANSFER next earliest scheduled event on the EVENT list to the CHRONOLOGICAL list. (Cross out entry on EVENT LIST).
	10. ADVANCE Sim Clock to time of transferred event.
	11. TERMINATE work tasks with associated E.E.T.
	12. RELEASE unit(s) to following elements.
	13. CHECK STAT clock switches (Same as step 4).
	14. Return to step 1.

The first activity to be commenced is work task 2, and the units are moved from QUEUE nodes 1 and 4. This would trip the laborer unit STAT clock to the OFF position, but this is not relevant, since it was never turned on. The time generated is 3.0 and the EET for COMBI 2 is 3.0. This is recorded, and the algorithm moves to the ADVANCE phase.

The EET of 3.0 is transferred to the CHRONOLOGIC list and the SIM TIME is advanced to 3.0. This results in the termination of COMBI 2 and the release of the laborer and stack unit. In its return transit to QUEUE node 1, the laborer passes the CLOCK ON position, the unit number, L1, is recorded, and the ON column checked.

Returning to the GENERATE phase, the COMBI 2 work task can be started again. The laborer unit moves to the LOAD SCAFFOLD work task, passing the CLOCK OFF position and tripping L1 STAT clock OFF. A check is placed in the OFF column, and the DURATION value is calculated as zero. This happens because the SIM TIME value has not advanced since the L1 STAT clock was switched on. Simulation of further cycles to observe the action of the STAT clocks is left as an exercise for the reader.

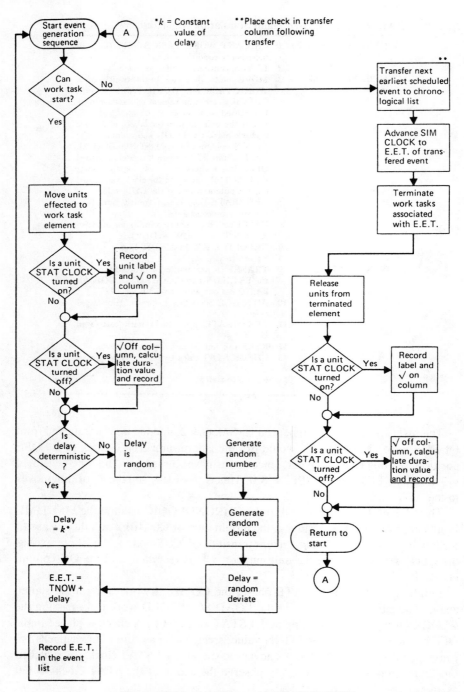

Figure 10.14 Algorithm with STAT CLOCKS.

11 Typical Repetitive Processes

11.1 INTRODUCTION

The preceding chapters have introduced the concepts of model building and simulation as a methodology for studying and analyzing construction operations. The next chapters will introduce the application of this methodology in the context of examples from actual field situations. This chapter will focus on various repetitive process models typical of construction activities. The following chapters will look at models from the building construction and engineering construction areas as well as other more specialized construction processes. All of these operations have the common trait that they are repetitive or cyclic in nature. The reader will note that in modeling complex real-world construction processes the materials handling aspects of the work often control the production rate and therefore greatly influence the structure of the final model.

The key to modeling operations to determine productivity and balance among resources is to identify processes that are repetitive. At the production level many processes are cyclic in nature and can be readily modeled using the CYCLONE modeling format. In general, processes that are linear or evidence linear characteristics are repetitive and good candidates to be modeled using a cyclic modeling environment. Some examples of repetitive or cyclic construction processes are

1. Concrete pouring
2. Structural steel erection
3. Slurry wall construction
4. Steel pile driving
5. Caisson foundation pouring
6. Pipe laying
7. Brick and masonry work
8. Reinforced earth construction
9. Exterior panel installation
10. Window or glass-curtain wall installation
11. Tunneling and tunnel excavation
12. Precast concrete member erection

Simplified models of some of these repetitive processes will be presented and discussed in this chapter.

11.2 CONCRETE POURING USING A CRANE AND BUCKET

Concrete is one of the most commonly used materials in construction. Its versatility and ease of placement has enabled it to hold its place in the market despite the development of more sophisticated materials. Concrete may be used in many different ways on a project to include monolithic foundation pours and precast panels to form the building exterior. Concrete can be either cast in place on-site or precast off-site and transported to the project for installation. The methods of placement at the site vary and depend on a number of considerations such as the (1) placement location (2) the desired speed of placement, and (3) the types of equipment available. In this section, a simple model for the placement of transit mix concrete using a crane–bucket system will be discussed.

It is assumed that forms and steel reinforcement are in place and that the system is not constrained by the batch plant (i.e., the quantity of concrete available is not a constraint). Required resources for the model include concrete hauling trucks, a crane with bucket(s), vibrating and finishing apparatus, and a crew of laborers at the placement site.

To provide a context for this model, assume that the concrete is to be placed on a paving job and that the concrete is batched at a site 2 mi from the paving site. Rather than arriving at the site as a wet mix, five-at-a-time dry batches are carried in an open bay truck (with appropriate compartments) to a mixer near the paving site. The batches are then dumped individually and sequentially into the skip of the mixer and mixed sequentially, starting with batch 1 and ending with batch 5. As each wet batch exits the mixer, it is dumped into a concrete bucket and lifted by a crane to the placement location where it is dumped, spread, vibrated, and finished by a concrete crew. A schematic diagram of the process is shown in Figure 11.1a. Although dry batching operations of this type are not common, this situation provides a good opportunity to utilize various modeling features. A picture of a crane and bucket placement operation is shown in Figure 11.1b.

A CYCLONE model of the process is shown in Figure 11.2. The model consists of six cycles representing the various flow units involved. The units and cycles of interest are

1. Batch plant
2. Trucks
3. Mixer
4. Crane
5. Bucket
6. Laborer crew for spreading, vibrating, and finishing

The process begins with trucks being loaded at the batch tower (COMBI 2). The trucks consist of five compartments, which are defined by baffles or dividers that are pinned in such a way that they can be released individually (one at a time) when the truck bed is elevated. Dry batches are loaded into each of the five compartments. This requires five individual loads at COMBI 2. The demand for five loads is generated using the GENERATE function at QUEUE node 9. When five batches are loaded, the CONSOLIDATE at FUNCTION node 3 assembles the five loads into a single truck for travel to the mixer. Upon arriving at the mixer, the truck is again reconfigured to represent the five dry batches using the GEN-ERATE function at QUEUE node 5. Each of the five dry batches are dumped sequentially into the skip of the mixer.

The mixer processes the batches in sequence and converts them into wet batches for transport to the placement site. The space in the mixer is represented by a flow unit at QUEUE node 13. If one unit is initialized at QUEUE node 13, this means that the mixer is a single-drum unit and only one batch at a time can be processed. If two units are defined at QUEUE node 13, then the mixer is a dual-drum unit and 2 batches can be processed simultaneously (in tandem).

Once a batch is dumped into the skip of the mixer, it is moved to the drum, where water is added and it is mixed (NORMAL 10). Following mixing it occupies space in the drum (QUEUE node 11) until it can be dumped to the concrete bucket. Therefore, the space in the drum is not free until the bucket is filled at COMBI 12 and this is represented by the feedback loop to QUEUE node 13 (the space QUEUE). That is, space becomes available after the bucket is filled. This avail-ability of space is required before the next dry batch can be loaded to the mixer at COMBI 6.

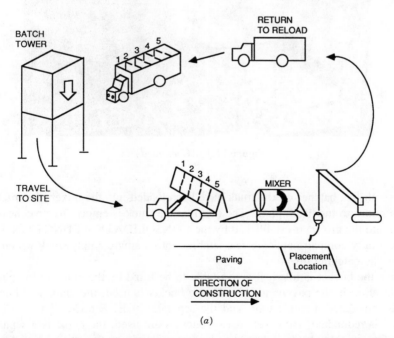

(a)

Figure 11.1   (a) Dry batch delivery and placement; (b) crane–bucket concrete placement.

Figure 11.1   (*Continued*)

Once all five batches on the truck have been loaded into the mixer, the truck is free to return to the batch plant. The fact that the truck is empty (five batches are loaded into the mixer) is established by the CONSOLIDATE at FUNCTION node 7. The empty compartments are reassembled into a single empty truck which returns to the batching tower.

When the bucket is filled, it is available to be lifted by the crane to the placement location in the pavement. If only one bucket is used, the crane and bucket can be considered a single unit, and the separate QUEUE node 17 for "Crane Ready" is redundant. However, if two buckets are used, the crane is a separate unit. After swinging back, it drops one bucket and picks up the other. This is more efficient since the bucket at the mixer can be filled while the crane is swinging and

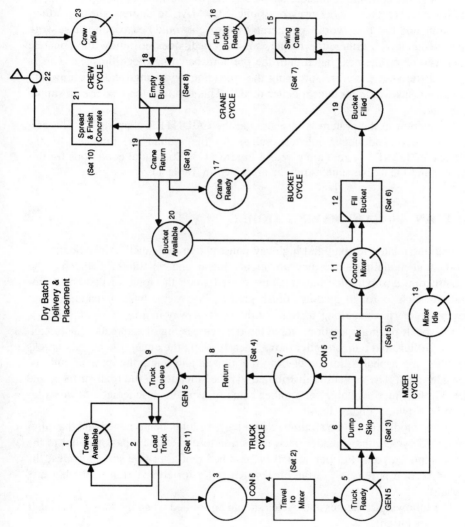

Figure 11.2 Dry batch and delivery placement model.

placing the other. The second bucket is a storage similar to that discussed in the previous chapter and keeps the crane active rather than tying it up during the "Fill Bucket" activity at 12. For this reason, the crane cycle and the bucket cycles are separated but nested. One cycle is nested inside the other.

Finally the concrete is placed, spread, vibrated, and finished in one activity at NORMAL 21. This model uses only one NORMAL to address the spreading, vibrating and finishing work. An alternate approach would be to define these work tasks separately. Further, the model, as structured, does not allow the concrete bucket to dump the next batch until the previous batch has been finished. This is not realistic and, therefore, breaking the spreading, vibrating, and finishing into separate tasks would produce a better model. This is left to the reader as a simple exercise.

Production in the system is measured at the COUNTER element 22. The production curve and queue idleness values for the system as simulated by the MicroCYCLONE* system are given in Figure 11.3. The initial conditions for this system and the active state durations are given in Table 11.1.

11.3 AN ASPHALT PAVING MODEL

Operations related to road and highway construction are excellent candidates for production modeling since they are highly linear and repetitive. Typically, the construction is subdivided into stations located along the route of the work. Operations such as rough grading, finish grading, aggregate base preparation, and paving are performed along sections of the right-of-way in a repetitive fashion.

In asphalt paving operations, a paving train consisting of a spreader, a "breakdown" roller, and a finish roller move linearly along the area to be paved. Trucks haul hot mix asphalt from the plant to the job site and dump the material into the spreader skip. The asphalt is distributed via the spreader to the road surface, and the skip becomes available for another batch of asphalt. A schematic of this situation is shown in Figure 11.4a.

In this model, it will be assumed that a parking lot is being paved and that after 15 spread cycles, the spreader must reposition to make a new pass parallel to the just-completed pass. Further, it will be assumed that after five spread cycles, the spread section is released to the breakdown roller for compaction of the hot mix asphalt.

The following resources and cycles should be studied when modeling an asphalt paving operation:

1. Spreader
2. Trucks
3. Breakdown roller
4. Finish roller
5. Asphalt plant

---

*The MicroCYCLONE computer system is described in Appendix C.

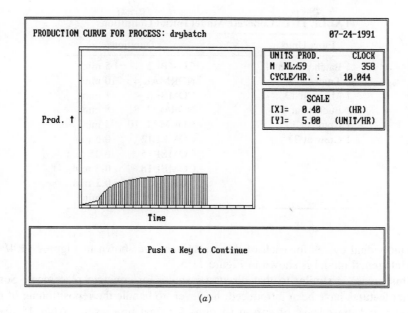

(a)

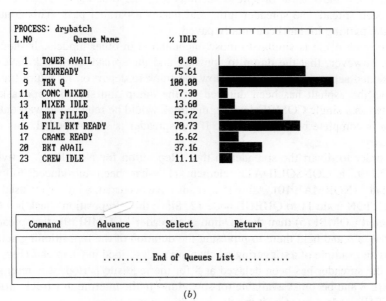

(b)

Figure 11.3 Concrete placement model: (*a*) production curve; (*b*) idleness.

TABLE 11.1   Concrete Model Initial Conditions

Flow Units	Durations	
1 Batch plant at 1	COMBI 2	5 min
4 Trucks at 9	NORMAL 4	10 min
1 Mixer at 13	COMBI 6	1 min
2 Buckets at 20	NORMAL 8	8 min
1 Crane at 17	NORMAL 10	3 min
1 Crew at 23	COMBI 12	0.5 min
	COMBI 15	0.25 min
	COMBI 18	0.3 min
	NORMAL 19	0.2 min
	COMBI 21	5 min

The individual cycles for each of these resources are shown in Figures 11.4*b–f*. The integrated model is shown in Figure 11.5.

This model is similar to that discussed above for concrete placement. Some special features have been introduced, however, to handle the repositioning of the spreader and the release of spread sections for final processing. After 15 loads have been spread, the spreader turns and makes a parallel pass. This continues until the parking lot has been totally paved.

The truck cycle is similar to those encountered in other models. It should be noted, however, that the dump to spreader and the spread work task have been represented separately. This is to allow the truck to depart on its return travel as soon as the asphalt has been dumped. If the dump and spread work tasks are modeled as a single COMBI element, the truck would be required to wait until the spread is complete before departing. If the spreader is self-propelled, this is not necessary.

In order to divert the spreader to the "Reposition for New Pass" activity at COMBI 9, a CONSOLIDATE element (11) has been introduced following "Spread" (NORMAL 10). After 15 spreads have occurred, a unit is released from FUNCTION node 11 to QUEUE node 12. Since the "Reposition" task is a lower numbered COMBI (5) than the "Dump to Spreader" COMBI (9), the spreader is diverted to 5 and held there to represent the duration of the repositioning activity. This is an example of a triggering mechanism as described in Chapter 7 (Fig. 7.8). Once the spreader has been delayed at 5 for the requisite period, it is returned to QUEUE 8 and becomes available for spreading. In the interim, any trucks arriving at QUEUE node 4 must wait for the repositioning to complete.

The requirement to release sections to the "breakdown" roller in five spread packages is implemented by using another CONSOLIDATE at FUNCTION node 13. Each time five spreads have been made, a unit is released to QUEUE node 14 representing a section ready for compaction. If the roller is available, it will start the compaction work. If it is busy with a previous section, the spread section waits until the roller is available. Once the section is compacted, it is moved to QUEUE

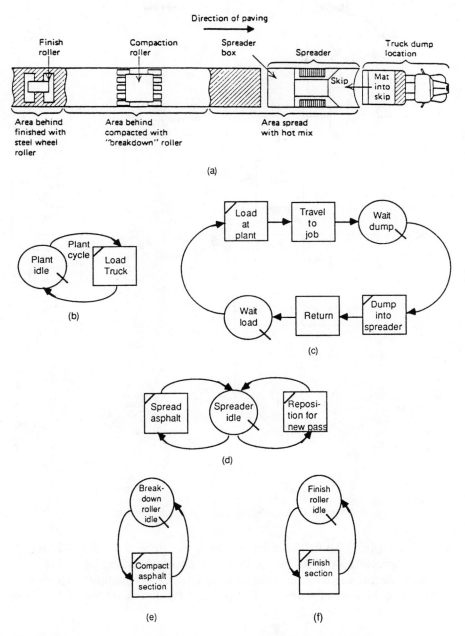

Figure 11.4 (*a*) Asphalt paving train; (*b*) plant cycle; (*c*) truck cycle; (*d*) spreader cycle; (*e*) "breakdown" roller cycle; (*f*) finish roller cycle.

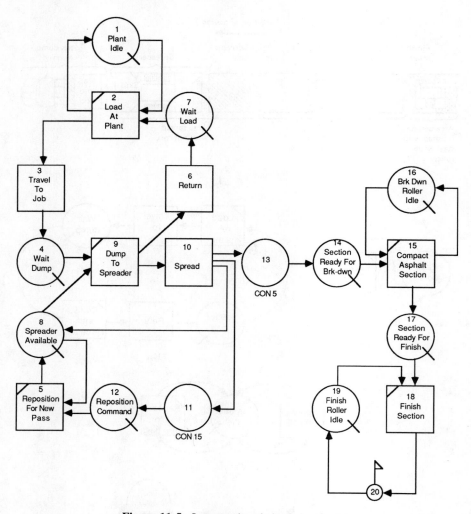

Figure 11.5   Integrated asphalt paving model.

node 17 and is ready for finish rolling. The units that arrive at the COUNTER (20) represent 5 truck loads of hot mix or 33.3% of a lane of parking lot paving.

The production curve for the model and the QUEUE node idleness information are shown in Figures 11.6 and 11.7. The initial conditions information for this model is given in Table 11.2.

11.4   A TUNNELING OPERATION

As an example of a more complex linear construction process, consider the tunneling operation shown schematically in Figure 11.8. In this process, precast concrete pipe is pushed forward using a jacking system. Penetration is achieved by

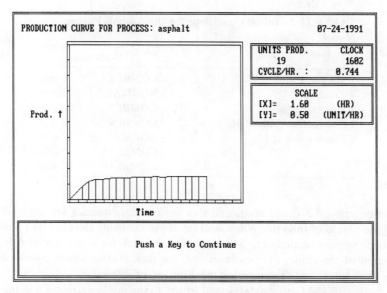

Figure 11.6 Asphalt paving operation: (*a*) production curve; (*b*) idleness.

excavating the tunnel face using a tunneling machine. The tunneling machine is jacked against the exposed edge of the nearest piece of precast circular pipe. Other jacks at the access shaft push the entire liner (consisting of individual pipe sections) forward.

Assume that the tunneling machine operates continuously 24 hr per day with a

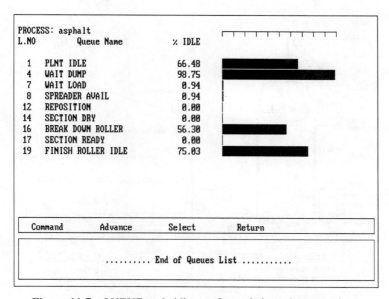

Figure 11.7 QUEUE node idleness for asphalt paving operation.

TABLE 11.2   Initial Conditions for the Asphalt Paving Model

		Durations
1 Plant at 1	COMBI 2	5 min
4 Trucks at 7	NORMAL 3	10 min
1 Spreader at 8	COMBI 5	22 min
1 Compaction roller at 16	COMBI 9	2 min
1 Finish roller at 19	NORMAL 10	12 min
	COMBI 15	35 min
	COMBI 18	20 min

production rate of 2 ft per hour, and that maintenance breaks are only allowed during the period of time that a new section of pipe is being lowered into position. The concrete pipe sections are 6 ft long, and the jack rams also travel 6 ft. The interaction of the crane, the pipe sections, the jack system at the bottom of the access shaft (at A) and the tunneling machine are of interest.

Suppose a model is to be developed of the tunneling operation for estimating the production in linear feet per day. The initial step in the modeling procedure is to identify the relevant system flow units. Certain of the units are obvious.

The following flow units seem to be good candidates:

1. Pipe sections
2. Crane (for lowering pipe sections)
3. Jack set

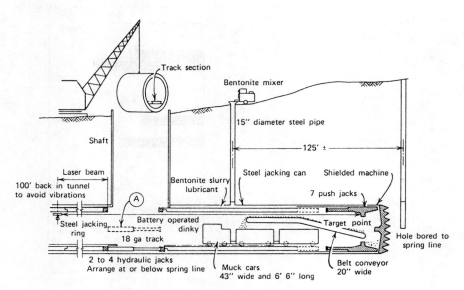

Figure 11.8 Tunneling process schematic.

4. Jacking collar

5. Crew for positioning pipe sections

It also appears that some type of informational unit will be required to link the pipe section lowering operation of the system with the jacking operation. The message to lower the next section of pipe is triggered by the jack rams having pushed the previous section as far as possible. When this occurs, an informational unit will be required to signal the crane to start the lowering pipe section sequence.

The units that are common to both lowering and jacking sequences are the pipe sections. These are the processed or mainline units that flow completely through the system. As such, their idle and active states indicate the positioning and definition of other resources required for the process. The level of detail specified in the definition of the processed unit cycle also establishes the total model detail and the complexity of the model.

The work tasks that are clearly associated with the pipe section flow are

1. Lower and position pipe.

2. Replace jacking collar and jacks.

3. Jack section forward 6 ft.

It is assumed that sections of precast concrete pipe are stockpiled at a location near the top of the access shaft. Following a message from the jacking operation that movement of the previous section has proceeded as far as possible (i.e., 6 ft), the crane maneuvers, picks up a new section and lowers it into position. A crew working in the tunnel helps to position the lowered section properly.

Following the positioning of the pipe, the jacking collars and the jack set are placed and prepared for further operation. When this is completed, the pipe section is jacked 6 ft forward, and the sequence repeats itself. For both activities, the work crew used for positioning is actively involved.

The cycle for the pipe section flow is shown in Figure 11.9a. The section is shown exiting the system into a queue or idle state. However, in accordance with the convention established previously, it is routed back to the pipe stockpile in order to achieve a closed path.

Having now developed the processed unit cycle, examine the cycles for resources required in moving the processed unit through the cycle. The crane cycle is simple, since the assumption is that its only work task is the lowering of the pipe into the access shaft. This means that it is "slaved" to this work task and passes through only two states, as shown in Figure 11.9b.

If the crane's work tasks included the off-loading and stacking of incoming pipe in the stockpile, the slave pattern would be expanded to a butterfly pattern. This revised listing of work tasks would lead to the cyclic flow shown in Figure 11.9c. The expanded model illustrates better and in more detail how the crane is actually committed. It does, however, imply the inclusion in the model of a truck cycle that interacts with the off-loading work task. For this example, consider that the crane's activity is limited to that of lowering the pipe sections.

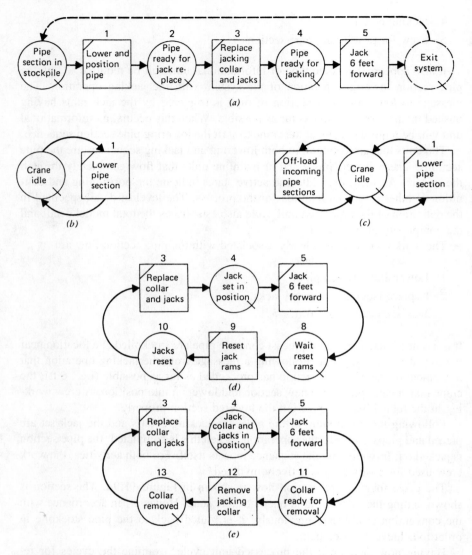

Figure 11.9 Model segments for tunneling process: (*a*) pipe section cycle; (*b*) crane cycle; (*c*) dual-task pattern for crane; (*d*) jack-set cycle; (*e*) collar cycle.

The jack set has two work tasks that are common to the pipe section cycle: (1) "Replace Collar and Jacks" and (2) "Jack 6 Ft Forward." These two work tasks must be included in the jack-set cycle. The question is: "Are there any additional work tasks to be considered?" Following the extension of the jack rams to their full length (6 ft), the rams must be reset. This provides a location for the insertion of the next pipe section. The resetting work task, however, is not part of the pipe cycle and must be added only to the jack cycle (see Fig. 11.9*d*).

The flow units that interact with the jack set cycle can be developed by examining the work tasks and the waiting states that precede them. The pipe sections are required for the "Replace" and "Jack" work tasks (work tasks 3 and 5). The jacking collar is replaced at the same time as the jacks, so that it is also required for all of the operations in this sequence.

The jacking collar cycle work tasks can be developed from the three cycles developed up to this point. The collar is required for both the "Replace" and "Jack" work tasks. The collar must also be removed from the jacked section in addition to being replaced on the newly inserted section. The removal of the collar is not included in any of the cycles developed previously. Therefore, it must be unique to the "Collar" cycle. The collar cycle based on these three work tasks appears as shown in Figure 11.9e. Again the waiting states indicate the array of resources that are required for the "Collar" to negotiate its cycle. The pipe sections, the jack sets, and the jacking crew constrain the flow of the collar through its cycle.

Having now developed four cycles, a tentative model can be assembled by integrating the common work tasks. These work tasks represent points at which the cycles are tied together (i.e., points at which they interface).

The four cycles lead to the specification of five work tasks that define the columns of a composite ingredience matrix. This matrix is shown in Table 11.3.
The flow units selected define the rows of the matrix. Of those units defined, only the crew flow cycles have not yet been investigated. Two crews have been defined to simplify the structure of the initial model for illustrative purposes. However, following initial structuring, an alternate model with only one crew performing all operations can be developed.

By examining the columns of the composite ingredients table, one can determine the paths that flow through each work task and, consequently, the waiting states that precede the work task. The idle states (i.e., QUEUE nodes) that precede a given work task are uniquely associated with it and they constitute the ingredients set required to allow commencement of the work task. Thus three elements—(1) pipe sections, (2) crane, and (3) crew 1—are required for the commencement of

TABLE 11.3 Composite Ingredients Table

	Lower Pipe (1)	Replace Jacks and Collar (3)	Jack 6 ft Forward (5)	Reset Jack Rams (9)	Remove Jacking Collar (2)
Pipe section	X	X	X		
Crane	X				
Jack set		X	X	X	
Jack collar		X	X		X
Crew 1	X	X			X
Crew 2			X	X	

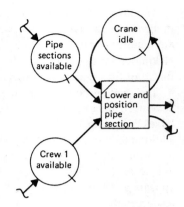

Figure 11.10   Network segment for work task 1.

the work task 1, "Lower Pipe." In graphical notation, this requirement results in a model segment, as shown in Figure 11.10.

The composite ingredients table enables the ingredients set associated with a work task to be determined by reading down the columns. By reading across each row, the work tasks through which a unit must flow can be determined. The network segment for each column in the composite ingredients table and the total composite (integrated model) are shown in Figures 11.11 and 11.12. Since the flow cycles for the crews have not yet been defined, they are left to be completed.

Three of the flow paths exiting work task 3 pass on to work task 5. In order to maintain the preceding idle state logic developed in Figure 11.11c, three separate idle states (one for each of the flow units) are shown. The three idle states can be consolidated into one QUEUE node. That is, the three flow units are considered to be combined for the transit between work tasks 3 and 5. This leads to a less cluttered graphical model, as shown in Figure 11.13. Since the pipe, jacks, and collar have been combined at 3 and flow together to work task 5, "Jack," the ingredients set can be represented by a single unit instead of by three separate units.

11.5   CREW UNIT FLOW PATTERNS

At this point a decision must be made concerning the flow of the crew units. Examination of the crew 1 row of the composite ingredients table (Table 11.3) indicates that this crew unit is required for the commencement of three work tasks. Two of these three work tasks occur sequentially. Work task 3 follows 1 directly without another intervening active state. Work task 3 also follows 12 directly (i.e., without any intervening active states). Five flow patterns for CREW ONE are possible. These are shown in Table 11.4.

Obviously, pattern 1 would not be appropriate to the tunneling operation; as long as there are pipe sections in the stockpile, the crew 1 unit will be continuously rerouted back to COMBI element 1, since priority is given to lowest-numbered

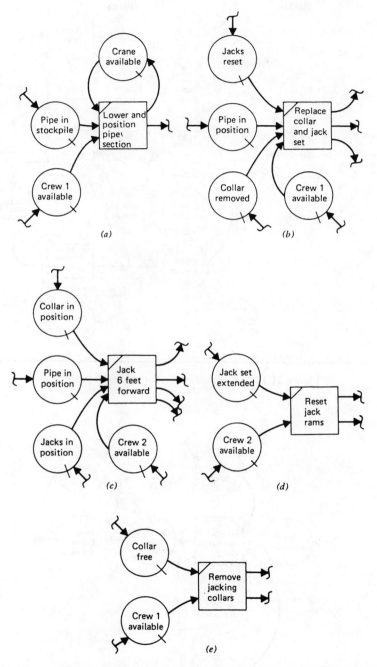

Figure 11.11 Network segments: (*a*) lower pipe; (*b*) replace jacks and collar; (*c*) jack 6 ft forward; (*d*) reset jack rams; (*e*) remove jack collar.

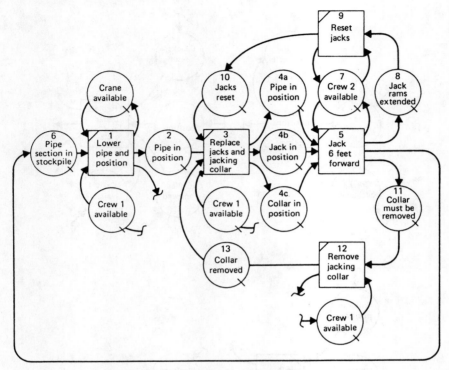

Figure 11.12   Integrated model for modeling process.

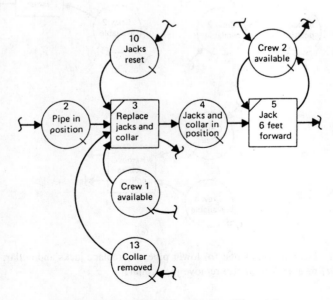

Figure 11.13   Consolidated unit waiting state.

TABLE 11.4 Possible Crew 1 Flow Patterns

	Flow Pattern	Implied Constraints	Graphical Form
1	1, 3, 12	Unit returns to be rerouted after each active state. No topological constraints on movement. Only constraints established by labeling.	
2	1, 3 → 12	Caught in loop 3 → 12 Implies units at 1 must wait if unit is between 3 → 12, since it can only be rerouted to 1 after passing 12 and returning to the idle state (16).	
3	1 → 3, 12	Caught in loop 1 → 3 Implies 12 must wait if unit is between 1 and 3.	
4	1 → 12, 3	Caught in loop 12 → 1 Implies units at 3 must wait if crew unit is between 12 → 1.	
5	1 → 3 → 12	Constrained by sequential flow pattern. Between 3 → 12 can't go to 3 or 1. Between 12 → 1 can't go to 12 or 3. Between 1 → 3 can't go to 1 or 12.	

211

work tasks. This would lead to a continuous lowering of pipe sections into the access shaft with no forward tunneling. Having piled all the pipe sections in the access shaft, the crew could quit and go home. A situation would result similar to the action of the wizard's broom in the well-known fairy tale about the sorcerer's apprentice.

Flow pattern 1 is desirable from the modeling standpoint, since it requires the addition of only one modeling element, the idle state 16. However, in order to remedy the problem described above, a control unit must be added to keep the crew unit from continuously cycling back to work task 1 (i.e., from continuously lowering pipe sections). This can be achieved by defining a message or informational flow unit that is sent from the "Reset Jack Rams" work task (9) to the "Lower Pipe and Position" work task (work task 1). This leads to the following ingredients requirements for work task 1:

	Pipe	Crane	Jack	Collar	Crew	Message
Lower and position pipe section	Yes	Yes	No	No	Yes	Yes

The revised model structure to include the newly defined informational flow unit is shown in Figure 11.14.

The definition of the message unit leads to the addition of a new idle state (17) to indicate the informational ingredience requirement at work task 1. Therefore two elements must be added to control the flow of the pipe sections through the system. The addition of the message flow cycle leads to an economy, however, since its path from 9 to 17 renders the idle state at 10 superfluous. The "Reset Jack" QUEUE node completing the jack cycle is nested inside the newly defined message cycle. Therefore, idle state 10 constraining 3 is not needed. Element 3 is constrained through work task 1 by idle state 17. Looking at it from a slightly different point of view, it is obvious that work task 1 requires a pipe section and a message unit. If only one message unit is defined (as would be the case in this instance), this constrains the number of pipe sections that can be flowing between work tasks 1 and 5 at any given time to one section. As previously modeled, the commencement of work task 3 requires a pipe section, the jack unit, which indicates that rams have been reset (10), and a crew. Since, the jack unit becomes immediately available following 9, as does the message unit, its presence is implicit if a message unit is available at 17.

The net effect of these additions and deletions is that the model has been increased in size by one element. Idle state 10 can now be deleted. Therefore the net addition is one element. A further economy can be realized if it is assumed that the requirement for crew 2 is not constraining and that it is always available at work tasks 5 and 9, as required. Because of the sequential nature of work tasks 5 and 9, and because the same unit processed at 5 is then processed at 9, this constraint is superfluous as long as only one pipe section is processed at a time. This is ensured by the message cycle, as discussed above. The deletion of these

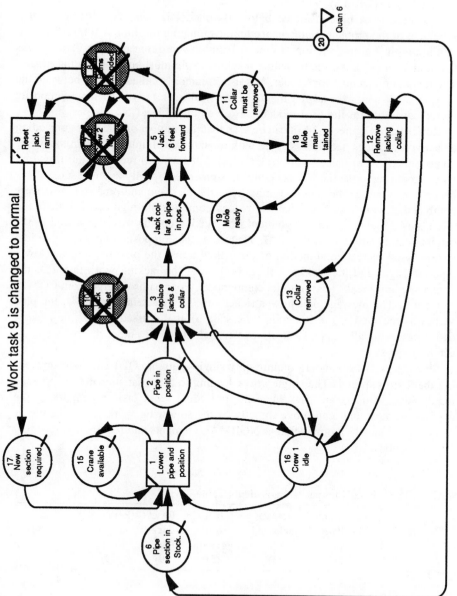

Figure 11.14 Total system with informational unit.

Work task 9 is changed to normal

Quan 6

idle states associated with crew two and the jack set cycles is indicated in Figure 11.14 by shading over these idle states 7, 8, and 10. Work task 9 can now be modeled as a NORMAL work task.

One item must be considered before the model is complete. In the original statement of the problem maintenance on the tunneling machine will be performed during the time a new section of pipe is being lowered into position. This means that a unit representing the tunneling machine cycle must be added to the model. The tunneling machine (or "mole," as it is commonly called) cycles between the jacking work task (5) and the maintenance work task (18). The new composite ingredients table for the final model is as given in Table 11.5.

Referring to Figure 11.14, it can be seen that the number of idle states preceding an active state does not always agree with the number of units shown in the ingredients table. For instance, work task 5 (jack 6 ft forward) requires four flow unit types according to the ingredients table. It is preceded by only two idle states. This occurs since two of the required units (i.e., the message and pipe sections) were combined at work task 1 and have remained combined through active state 3 and idle states 2 and 4. A third ingredient (the jacking collar) was also added to the original combination at work task 3. Therefore, the composite unit waiting at idle state 4 is, in fact, a combination of three flow units: the pipe section, the lower pipe message, and the jacking collar. Following the jacking work task (5), the composite unit breaks apart into its components. Four paths diverge from COMBI element 5. The message unit moves to activity 9 ("Reset Jack Rams"), the pipe section is recycled to the stockpile (idle state 6) as discussed above, the jacking collar passes to idle state 11 in its cycle, and the mole enters the maintenance activity.

Since work tasks 9 and 18 require only single units, no QUEUE nodes preceding them are required. The single arrow leading to them satisfies the ingredience requirement. This means that when the preceding work task is completed, the appropriate flow unit can move directly to the following work task. Therefore, work tasks 9 and 18 are modeled as NORMAL elements and require no preceding QUEUE nodes.

TABLE 11.5 Final Composite Ingredients Table

	Lower and Position Pipe (1)	Replace Jacks and Collar (3)	Jack 6 ft Forward (5)	Reset Jack Rams (9)	Remove Jacking Collar (12)	Mole Maintenance (18)
Pipe section	X	X	X			
Crane	X					
Jacking collar		X	X		X	
Crew	X	X			X	
Message	X	X	X	X		
Mole			X			X

The solution presented for the tunneling problem uses the message informational unit to control the lowering of new pipe section into the excavation. This causes the definition of an additional flow unit. It is possible to control the flow of pipe sections into the tunnel by utilizing flow pattern 5 from Table 11.4. This utilizes the CREW flow unit as the control unit, thus reducing the total number of units defined. Design of this system as well as the explanation of flow patterns 2 to 4 for the crew unit are left as exercises for the reader.

12 Building Construction Models

12.1 INTRODUCTION

Building construction processes are complex and require the coordination of a wide variety of crafts and technologies. Construction of multistory buildings results in repetitive processes since each floor of the construction can be thought of as a section or repeated phase of the work. Designers often optimize the design of buildings by capitalizing on modularity and repeated features in each floor. Tall buildings have the linear characteristic of repeated floors, which result in cyclic operations. In addition to the erection of the structural frame, the exterior cladding or finish of the building requires stone, precast concrete, or glass panels. The installation of this exterior "skin" of the building is highly repetitive and well adapted to production modeling. Other construction features such as forming and casting concrete are repetitive. This chapter will introduce some production models designed to aid in the analysis of building construction projects.

12.2 A SIMPLE CONCRETING MODEL

As previously noted, the placement of concrete is one of the most common construction processes encountered on any building construction job site. Particularly, in the construction of mid- to high-rise buildings, the rate of movement of the concrete from ground level to the location of placement controls productivity. The concrete is often produced at off-site batching facilities and moved to the site by transit mix trucks. On arrival at the job site the fresh concrete is moved to placement location by a crane, a hoist, or a pump. A variety of models can be constructed to analyze the rate of production. A simple model for the placement of floor slab concrete in a high-rise building is presented in Figure 12.1. In this particular model it is assumed that concrete is moved to the work floor by hoist, held temporarily in a storage hopper, and then distributed using rubber tired buggies. This model consists of four cycles that are linked together. The four cycles of interest are the (1) transit mix truck cycle, (2) hoist cycle, (3) hopper cycle, and (4) buggy cycle.

This model must use CON–GEN combinations as discussed in Section 7.3 to subdivide truck size batches into buggy-size units for placement. It is assumed that each truck represents 10 yd^3 of concrete (five lift or hoist loads) and that the arrival of each truck generates five cycles of the hoist. This means that the hoist capacity

216

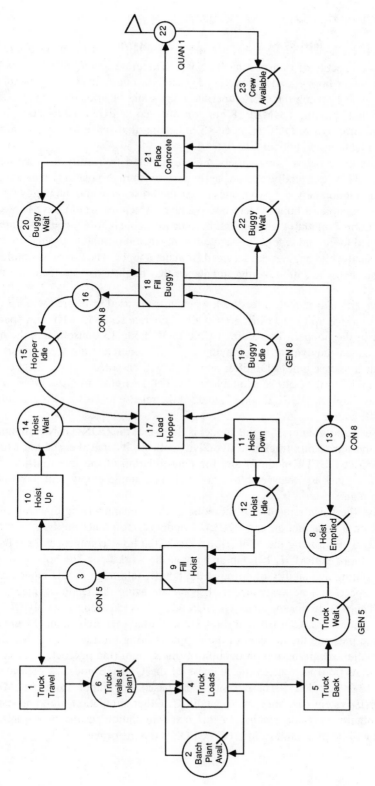

Figure 12.1 Model of concrete slab placement.

217

is 2 yd^3. Each 2-yd^3 volume is held in the storage hopper at the work floor. The buggies are capable of transporting 0.25 yd^3. Therefore, each hoist–hopper load represents eight buggy loads or cycles. These eight buggy loads are generated at QUEUE node 19 using the GEN function. Once eight buggies have been loaded, the eight units passing COMBI 18 are consolidated at FUNCTION node 16. A single unit proceeds to QUEUE node 15 indicating that space for an additional 2 yd^3 is available in the hopper.

Hoppers or storage of this variety are encountered frequently in construction processes. They essentially operate as buffers or storage positions balancing or linking production between two cycles (e.g., the hoist cycle and the buggy cycle). If, in this example, a hopper were not available, the hoist would have to wait at the work floor level until eight buggies became available before it could be emptied and returned to ground level. This is inefficient since returning to ground could be accomplished while buggies are serviced from the hopper. The use of this buffering mechanism helps to minimize the imbalance that might exist between two interacting cycles.

In this particular model, a feedback mechanism is used to insure the hoist does not start to move up to the work floor with a concrete batch (2 yd^3) until space is available in the hopper. Therefore, FUNCTION node 13 consolidates the eight buggies loaded (equivalent to removing 2 yd^3 of concrete from the hopper) and then sends a release unit to QUEUE node 8. It will be noted that the hoist cannot be filled until (1) the hoist is available at 12, (2) the transit mix truck (i.e., concrete) is available at QUEUE node 7, and (3) the release just discussed is available at QUEUE node 8.

Another point of interest in this model is the use of the CON–GEN combination to break the transit mix truck into five 2-yd^3 units on its arrival at the hoist (using the GEN 5 at QUEUE node 7) and the consolidation of the five transits of the "Fill Hoist" task to "reassemble" the truck as a single empty unit to return to the batch plant.

Finally, the basic structure of the hoist cycle is characteristic of many lifting activities on a construction site. QUEUE node 12 represents the hoist empty at ground level. Following the "Fill Hoist" task, the hoist commences its travel to the work floor (NORMAL 10). On arriving at the work floor, the hoist may have to wait pending availability of space at the hopper (although in this model provision that space will be available has been made using the feedback mechanism discussed above). In this example, QUEUE node 14 is required since only QUEUE nodes can precede a COMBI and the "Hopper Idle" QUEUE node 15 must be included to measure the idleness of the hopper waiting for a new 2-yd^3 batch.

Information regarding the production of this system (the production curve and hourly production) as well as the idleness of the QUEUE nodes is presented in Figures 12.2 and 12.3. These results are based on output from the Micro-CYCLONE system described in Appendix C. Other simulation programs can be used to obtain the same results. In this text, simulation results of the systems modeled will be presented in MicroCYCLONE report format.

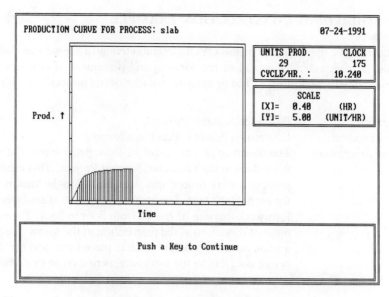

Figure 12.2 Concrete model production curve.

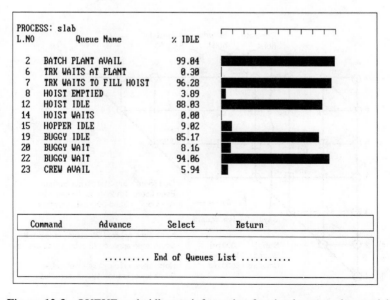

Figure 12.3 QUEUE node idleness information for simple concreting model.

12.3 A BUILDING CONSTRUCTION EXAMPLE

To illustrate the applicability of the CYCLONE modeling in a practical context, consider a more extended model of the forming and placement of concrete on a high-rise construction project. The general description of the project is as follows:

Project title	Peachtree Summit Building
Project location	Downtown Atlanta, Peachtree Street
Process description	The observed process involves the repetitive installation of floor slabs in the Peachtree Summit Project. This repetitive process occurs from floors 4 to 30. A single form is used for each 34-ft^2 bay. This results in the use of prefabricated formwork that can be reused from floor to floor. The movement of these forms, the preparation of the forms, the preparation of the forms for concrete placement, and the delivery of concrete to the forms are examined in this analysis.

Drawings showing the site layout and appropriate elevations are given in Figures 12.4 and 12.5.

The steps involved in the forming and concreting process are as follows:

1. A 2-yd^3 concrete hoist is fed by a 10-yd^3 concrete truck. Hoist cycles are required to empty the truck.

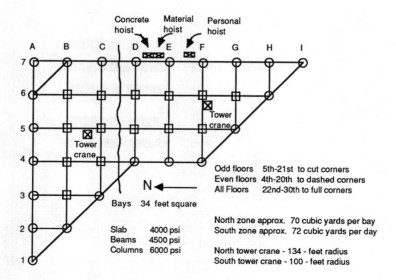

Figure 12.4   Site layout.

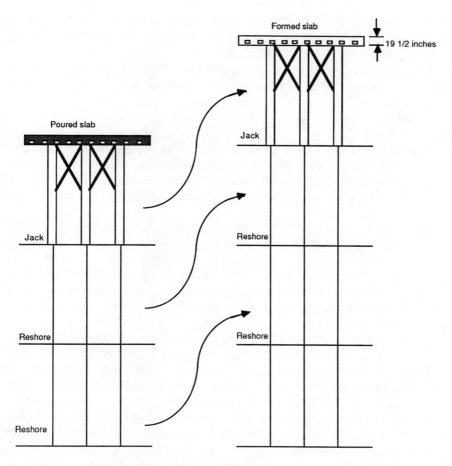

Note: Slab form keeps its same relative position on each floor, fits
      into slots on slab

Figure 12.5 Floor slab profile.

2. The concrete hoist is lifted to the floor level under construction. A 2-yd^3
    hopper is filled with concrete from the hoist. The hoist returns to the ground
    for another load from the truck.

3. A 0.25-yd^3 buggy is filled from the hopper. Eight buggies are required to
    empty the hopper.

4. The buggies carry concrete to the formed bay, where it is poured and fin-
    ished.

5. After curing has taken place, forms are "flown" to the next floor using the
    crane. Forms are leveled and prepared for the next pour.

This process is shown in CYCLONE system format in Figure 12.6. It introduces several interesting requirements for the use of GEN–CON combinations as described in Chapter 7 (Section 7.3).

First, the arriving concrete truck must be broken into five hoist loads. This is accomplished using a GEN–CON combination with an N value of 5 at elements 4, 5, and 6. That is, five units are generated and then consolidated in the sequence 4, 5, and 6. This results in five cycles of the hoist cycle for each single cycle of the truck cycle.

Each hoist load (2 yd^3) is lifted to a temporary storage hopper. Here each 2-yd^3 hopper load is further broken into eight individual buggy loads, which are carried to the pour site. This subdivision is also handled using a GEN–CON combination. At this point we have converted the original truck load of concrete into 40 (5 × 8) buggy loads.

Since each bay pour takes approximately 70 yd^3 of concrete, the total bay pour requires 280 buggy loads (70 yd^3 divided by 0.25 yd^3 per buggy). Certain actions must be initiated when the pour is complete. In order to note the completion of the "Concrete Pour," a CON 280 element is inserted following COMBI 21. Similarly, in order to generate the demand for 280 buggy loads once the forms are ready for pour commencement, a GEN 280 is associated with element 29. This insures that the buggies will cycle 280 times.

Certain control mechanisms have been inserted to ensure that the concrete delivery is coordinated with demand. A feedback loop is inserted from element 18 to COMBI 5. This ensures that a concrete truck does not begin to unload until space is available in the hopper. This action is implemented by consolidating eight buggy loads at element 7 and then issuing an "Unload Truck" command at 8.

Furthermore, the trucks on the haul cycle must be stopped once the pour is complete (280 buggy loads are placed). Otherwise they will continue to haul, filling the system with concrete that is not required and must be wasted. Depending on how many trucks are on the haul, this command to stop hauling must be given to a certain number of truck loads prior to completion of the slab. For simplicity, assume that the command is released after 280 buggy loads.

Implementation of the "Stop Order" is achieved using a capture mechanism that withdraws the load permit associated with COMBI 15. That is, when the 280 cycles have been completed FUNCTION node 22 sends a unit to QUEUE node 31. This means that ingredience conditions at COMBI 2 are met as soon as the "Load Permit" is available at QUEUE node 1. Even if trucks are available at QUEUE node 3, the "Load Permit" will be forced to detour to COMBI 2 ("Capture Load Permit") rather than returning to COMBI 15 since the lower-numbered COMBI controls the flow from QUEUE node 1.

In order to check this system for logical integrity, a simulation can be conducted. Deterministic durations will be used for simplicity. Basic data for the system simulation are given in Table 12.1.

Various simulation techniques can be used to evaluate the production and QUEUE node idleness. The production curve (buildup of production over time) and information regarding QUEUE node idleness generated by the Micro-

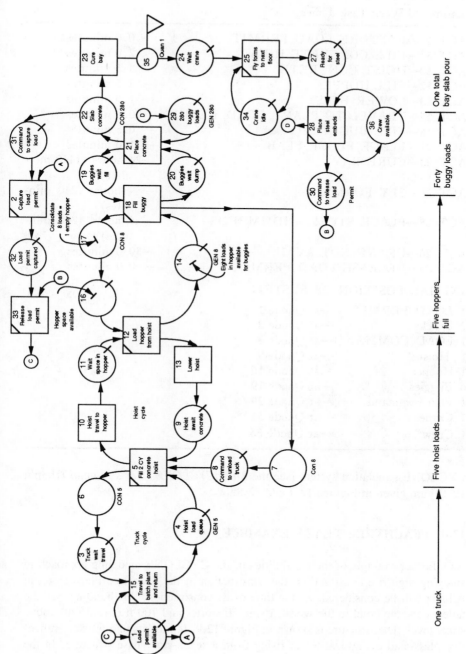

Figure 12.6 Building construction process model.

223

TABLE 12.1   Basic System Data

Estimated Work Task Times

ACT 2 —CAPTURE LOAD PERMIT	— 0.0 minutes
ACT 5 —FILL CONCRETE HOIST	— 2.0 minutes
ACT 10—HOIST CONCRETE	— 1.5 minutes
ACT 12—FILL HOPPER	— 2.5 minutes
ACT 13—LOWER HOIST	— 1.0 minutes
ACT 15—TRAVEL TO PLANT, LOAD, RETURN	—20.0 minutes
ACT 18—FILL BUGGIES (1 each)	— 0.5 minutes
ACT 21—PLACE FLOOR SLAB	— 1.0 minutes
ACT 23—CURE BAY	—24 hours (1440 minutes)
ACT 25—FLY FORMS	— 4 hours (240 minutes)
ACT 28—PLACE STEEL AND IMBEDS	— 4 hours (240 minutes)
ACT 26—CRANE NOT AVAIL	—30.0 minutes
ACT 33—RELEASE LOAD PERMIT	— 0.0 minutes

INITIAL POSITION OF SYSTEM

1 LOAD PERMIT	—at Qnode 1
2 Trucks	—at Qnode 3
1 LOAD COMMAND	—at Qnode 8
1 Hoist	—at Qnode 9
1 Hopper	—at Qnode 16
6 Buggies	—at Qnode 19
1 Pour command	—at Qnode 29
1 Crane	—at Qnode 34
1 Crew	—at Qnode 36

CYCLONE simulation system [see the *MicroCYCLONE User's Manual* (Halpin 1990)] are given in Figures 12.7 and 12.8.

12.4   PEACHTREE PLAZA EXAMPLE

As a further example of the capabilities of the CYCLONE modeling approach in studying high-rise construction, the construction of the Peachtree Plaza Hotel in Atlanta will be considered. At the time of its construction, this building was the tallest concrete hotel in the world, rising 70 stories and 700 ft (213.36 m) above street level. The structure is shown in Figure 12.9. It consists of a 63-story reflective glass-clad cylindrical tower rising from a seven-story base structure. In the center of the base structure there is a seven-story central court or atrium, open from the lobby to a massive skylight at the base.

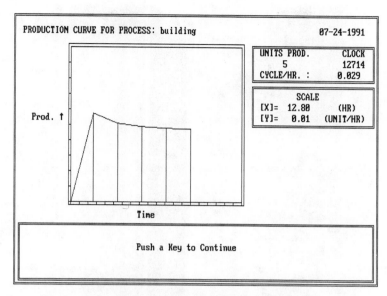

Figure 12.7 Production curve for Peachtree Summit process.

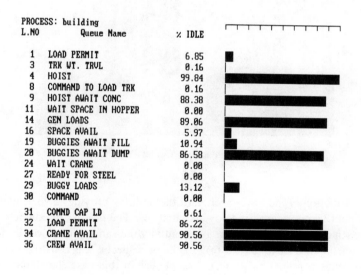

Figure 12.8 QUEUE node idleness statistics for building construction process model.

Figure 12.9 Peachtree Center Plaza Hotel.

The construction of the tower section of the building provides a very interesting linearly repetitive operation for study and analysis. The process of concrete placement for each of the 63 tower floors is essentially the same. The tower is slender, with a diameter of 116 ft (35.36 m). In many respects, the construction of the tower resembles a tunneling project, except that in this case the tunnel is constructed skyward along a vertical instead of a horizontal axis.

The model to be developed will focus on the concreting operation. The total job site is extremely congested, covering only 1.3 acres. Lifting of materials on the tower is handled by two tower cranes. The location of these cranes is shown in Figure 12.10. The cranes are "jumped" about once a week to keep up with the changing height of the structure. The jumping of each crane occurs during the weekend or nights when there is little or no work scheduled. The right crane on the tower is the smallest on site, with a radius of 80 ft (24.38 m). The left side or

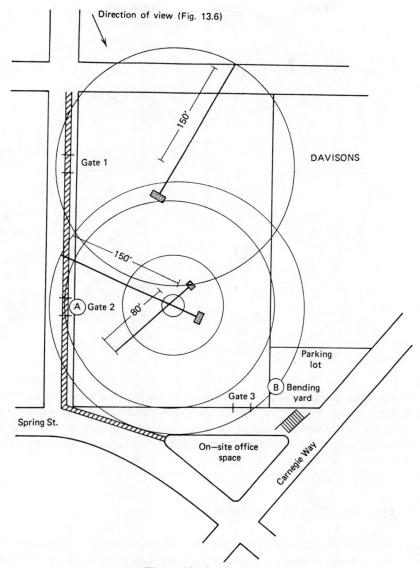

Figure 12.10 Site layout.

south crane has a 150-ft (45.72-m) boom length, allowing it to reach out into the street for direct off-loading of vehicles. A concrete hoist is used to move concrete from transit mix trucks at street level to the level of placement. A worker hoist is used for movement of personnel.

The layout of the job site in plan view is shown in Figure 12.10. A lane of the bordering street has been allocated for the queueing of concrete transit mix trucks (*A*). A bending yard for reinforcing steel is located at *B* on the roof of a parking garage next to the job site.

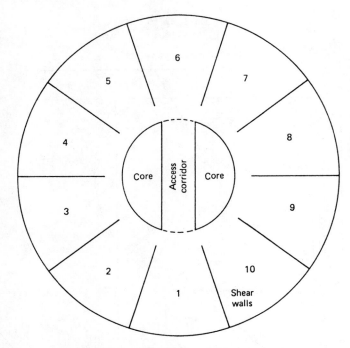

Figure 12.11   Typical floor section—plan view.

Disregarding special features, the sequence of concrete work for each of the 63 tower floors is repetitive. Each floor consists of a round slab with 10 shear walls radiating from lines that intersect at the center and a center core with walls at its boundaries. There is also a hall through the middle of the core that provides access to the elevators (see Fig. 12.11).

12.5   THE CONSTRUCTION PROCESS

The construction of each new floor begins with the telescoping or "extension" of the interior core form from the previous floor, as shown schematically in Figure 12.12a. The complete floor construction process consists of six steps that are repeated for each floor. After extension of the interior core forms, they are cleaned and prepared for concrete. At this time, also, the exterior core forms and the shear wall forms are pulled and lowered to the ground for cleaning.

Following this, wedge-shaped floor slab forms are lifted or "flown" from two floors below to be inserted between the shear walls of the just-completed floor. There are two sets of floor slab forms (i.e., 20 sections). This allows the set of forms on the floor immediately below to remain in place, providing temporary shoring. After removal of the slab forms from two stories below, temporary timber shores are inserted on that floor. After insertion and setting of the 10 sections of

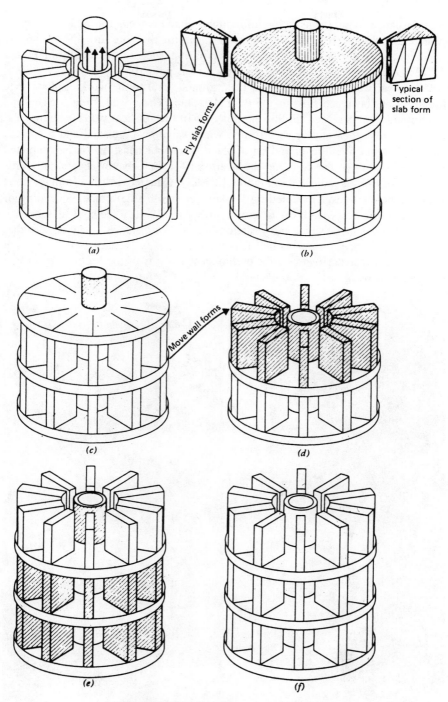

Figure 12.12 Concrete pour sequence.

wedge-shaped flying forms, the reinforcing steel and embeds for the floor slab are placed. In addition, the reinforcing steel cages for the core are lifted from the bending yard and placed in the core. This is indicated schematically in Figure 12.12*b*. After placement of steel, the floor slab is poured.

The shear wall forms are moved from ground level to the working floor once initial set has been achieved on the floor slab pour. These wall forms are set, and reinforcing steel is placed. Following this each of the 10 shear walls is poured and cured. Finally, the exterior core forms, which have been cleaned and prepared, are lifted from ground level and set in place. The floor construction is completed with the pouring of the core and the overhead for the elevator access hall. This sequence of tasks is indicated in Figures 12.12*c–f*. Concrete is lifted from the hoist with a concrete bucket mounted on the smaller (north-side) crane. Spreading of concrete at the working level is shown in Figure 12.13. The logical sequence described above is not rigid, and variations occur. For instance, it can be seen that the reinforcing steel for one of the shear walls is already in place as the floor slab pour is being made. However, for preliminary analysis of the process, it will be assumed that work tasks are sequential.

Figure 12.13 Spreading concrete.

12.6 DEVELOPMENT OF THE CYCLONE MODEL

The design of the process model of this operation requires the identification of a "mainline" unit of production as well as specification and allocation of the resources to be considered. The unit of production in this process is the completion of a floor of the hotel tower. Resources of interest include the two cranes used for materials handling on the tower and crews assigned to the placement of steel, the setting of forms, and the pouring of concrete. A model of the process is shown in Figure 12.14.

The model focuses on the activities developed in Section 12.5. The system boundary defined is such that the concrete hoist is not included. The lifting and placement of concrete is considered at work tasks 13, 22, and 27 for the floor slab, core, and shear walls, respectively. The lifting segment considered, however, is concerned only with movement of concrete from the hoist skip to the placement location. Therefore, in this model, the rate of the hoist is not assumed to be a constraining factor.

Certain assumptions have been made about the division of work tasks between the two cranes. The smaller crane is involved in the placement of the slab and core concrete, lowering of the core forms for cleaning, telescoping of the interior core forms, and lifting of floor slab steel. The larger crane is charged with the lifting of the flying forms and the placement of shear wall cages. It is also used in the placement of shear wall concrete and the lowering of shear wall forms to ground level for cleaning. Both exterior core forms and shear wall forms are lowered to ground level for cleaning because of the restricted working area on the tower itself. The work tasks are divided between the two cranes as follows:

Large Crane		Small Crane	
Act	Work Task	Act	Work Task
6	Fly slab forms	4	Telescope interior core form
11	Lift wall cages	10	Lift slab steel
27	Lift wall forms and pour concrete	17	Pull core forms
36	Pull wall forms	22	Place slab concrete

The lift priorities are established by the numbering sequence of the COMBIs. In lifting of steel, parallel operation is achieved by placing one crew–crane combination on each parallel track (e.g., 9–10–15–16 and 8–11–12).

Two types of crews are described in the model. Crew 1 is assigned work tasks related to slab steel placement, core form cleaning, and concrete placement in the core. Crew 2 is required for lifting and setting shear wall cages and shear wall removal of core and shear wall forms. Other units of interest in the model are the wall forms and exterior core forms that describe constraining cycles between 36–27 and 34–13, respectively.

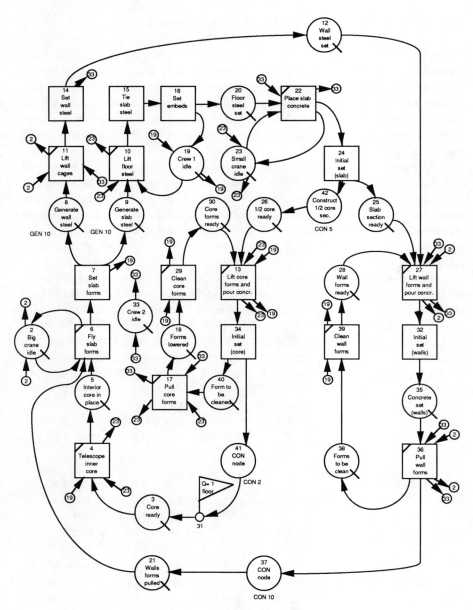

Figure 12.14 CYCLONE process model.

12.7 DEFINITION OF THE FLOOR CYCLE

It is informative to trace the flow of the mainline unit (the floor) commencing at element 3. The mainline unit subdivides and reconsolidates at various points in its cycle, leading to parallel tracks within the model. A unit is initialized at QUEUE node 8, indicating that initial set on the core has been achieved and that the floor cycle can commence. At COMBI 4, CREW 1, the small crane, and the ''Core Rdy (ready)'' message are combined, and the telescoping of the internal core form begins.

Following extension of the interior core form, the small crane is released and the slab forms from two floors below can be ''flown'' to the new level. This is contingent on the removal of all 10 shear wall forms from the level just completed. If all forms have been removed, the permit at element 37 will be available to transit to QUEUE node 21. Implicit in the commencement of the ''Fly Slab Forms'' work task is the availability of the big crane (QUEUE node 2), the removal of all shear wall forms (QUEUE node 21), the extension of the interior core forms, and the availability of CREW 1 (QUEUE node 5). Following the lift of all floor slab forms, they are leveled and prepared using CREW 1.

After installation of the slab forms, two work task sequences can commence. These sequences are initiated at QUEUE nodes 8 and 9. Ten flow units are generated into each sequence; they represent the breakout of each major physical component (i.e., slab and shear walls) into 10 subunits for processing. The slab is divided into 10 subsections corresponding to the 10 slab pours (sectors). The individual shear walls are also generated. The large crane is used for the lifting tasks only at elements 6 and 11. Therefore, it becomes available for other lifting tasks, while the crews are completing placement and tie-off of the steel. As noted, priority is given to slab steel lifts. When all steel and embeds have been placed in the slab, the pour of the floor proceeds, using the small crane and CREW 2.

After the pour and initial set of the floor slab, the two parallel steel placement sequences are converted into two new parallel tracks; one relates to the pouring of shear wall concrete, and the other pertains to placement of core concrete. Because of the breakout into subunits, the forming of shear walls and placement of concrete can proceed, even if only three or four shear wall reinforcing cages are installed. The allocation of cranes allows the large crane to continue to work on shear wall form installation, while the small crane handles core form setup.

At element 42, 10 units exiting the ''Initial Set (SLAB)'' task and representing the 10 slab pours (and associated core steel) are consolidated into two subunits that represent the two halves of the floor. The two halves of the floor release the half sections of the cores, indicating that reinforcing steel has been installed and the half core can be poured. The lifting and setting of the exterior core forms take priority over placement of wall steel in the model as configured. This allows parallel activity by the large crane in placing wall forms at the same time that the small crane sets a half section of the exterior core forms. This scheme avoids delaying installation of the exterior core forms until all shear wall steel has been placed.

After placement of the exterior core forms, the core is poured in half sections. In order to be compatible with the reduction in flow units from 10 to 2 (implemented by the CON 5 at element 42), two exterior core form units must be initialized at QUEUE node 30. Since the 10 shear walls are handled individually, no reduction is utilized prior to FUNCTION node 37. The shear wall ''Form and Pour'' track (27–36) is constrained by the number of prefabricated shear wall forms available. Since the forms must be pulled and lowered to ground level, a number of wall forms less than 10 leads to a constraint. After placement of concrete in the core, the exterior forms are also pulled and lowered to ground level for cleaning because of the extremely tight working conditions on the tower itself.

The floor cycle is completed with the consolidation of the two core half sections at 41 and the 10 shear wall form removals consolidated at 37. These two FUNC-TION nodes release permits required to commence the cycle. Production in floors completed is measured by the COUNTER inserted at 31.

12.8   MODEL INPUT AND RESPONSE

In the model as described, units are initialized as indicated in the following tabular list:

Number of Units	Unit Type	At Element
1	"Interior Core Form"	3
1	"Large Crane"	2
1	"Set of Wall Forms"	21
1 (or 2)	"Crew 1"	19
1	"Small Crane"	23
1 (or 2)	"Crew 2"	33
≤ 10	"Wall Forms"	28
2	"Ext. Core Forms"	30

The cycle time response of the system is indicated in Table 12.2.

The first item of interest to the construction manager is the question "How long will it take to complete one floor of construction?" The cycle times for the floor construction under varying conditions are presented in Table 12.2. This table is organized as a 4 × 4 matrix; the rows indicate variations of the crew combinations and columns indicate the number of shear wall forms available (range 10 to 2 forms). As would be expected, cycle times increase as the number of forms is reduced, causing a constraint in the segment 27–32–35–36. However, when using only one CREW 1 type unit (rows 1 and 3), the reduction to eight forms does not affect cycle time, indicating that no constraint occurs in the shear wall concreting segment. This obviously indicates that eight forms are sufficient to handle the input rate from element 24, provided only one CREW 1 unit is active in the slab steel area. When the number of CREW 1-type units is increased to two, the input rate increases, causing the constraint of only eight forms to become operative.

Examination of the 10-form column indicates that addition of an extra crew at 19 (i.e., increasing CREW 1 units to two) yields a cycle time improvement of 878 min or approximately 15 hr. Assuming that a double shift operation results in 15 hr of productive effort per day, this represents a time saving of approximately 1 day, reducing the cycle time from 5 days to 4 days. By contrast, increase of the number of CREW 2 units does not yield a significant reduction in cycle time (see row 3 in Table 12.2). This indicates that the system is relatively insensitive to

TABLE 12.2 Floor Cycle Times (in Minutes)

Number of CREW 1 \ Number of CREW 2	Ten Shear Wall Forms	Eight Shear Wall Forms	Five Shear Wall Forms	Two Shear Wall Forms
1 \ 1	4471	4471	4578	
2 \ 1	3593	3879	4260	6505
1 \ 2	4448	4448	4534	
2 \ 2	3374	3432	3764	6078

variation of CREW 2-type units. This is further verified by considering the 2 × 2 configuration of row 4 and comparing it to the 2 × 1 system of row 2. With 10 shear wall forms, the improvement in cycle time achieved by adding the second CREW 2 unit is (3593 − 3374) minutes or approximately 3.5 hr. On a cost-effectiveness basis, this addition normally would not be justified.

13 Heavy-Construction Models

13.1 REPETITIVE NATURE OF ENGINEERING CONSTRUCTION

Engineering or heavy-construction operations relate to the construction of roads, highways, airfields, subsurface structures, dams and other facilities which would not normally be considered buildings. Such operations require processes that are often very repetitive in nature. The earth hauling process has already been discussed in detail. The cyclic nature of road construction and asphalt paving has also been described in Chapter 11. Installation of storm- and wastewater sewer lines in urban areas is very repetitive, as is the construction of large pipelines for gas and oil (e.g., the Alaska pipeline). This chapter presents models that are representative of heavy construction operations.

Pipeline operations are similar to road and highway construction in that pipelines represent a linear work site that is broken into sections (stations) and worked on a repetitive (section-by-section) basis. The method of installing or laying pipeline varies with the size of the pipe, its location (e.g., above ground or subsurface), the method of connecting sections, and the internal and external pipeline pressures and integrity requirements, as well the composition of the pipe itself.

Typically, installation of large subsurface pipes requires excavation of a trench, the placement of a bearing–filter material to support the pipe, the placement of the pipe to appropriate elevation (invert level), and the backfilling of the trench. Open-trench methods are being replaced to some degree by "trenchless excavation construction" (TEC) methods such as pipe jacking. Pipe jacking is conceptually similar to the tunneling example in Chapter 11. The details of the approach vary based on the size of the pipe and the nature of the material being penetrated.

13.2 SEWER-LINE CONSTRUCTION MODEL

To demonstrate the modeling of a traditional "cut and cover" pipeline installation, assume that a sewer line is being constructed beneath the route of an existing city street. In this situation, a train of activities develops along the line of the installation. The major resources and work tasks are as follows:

1. *Pavement Removal.* A pavement breaker proceeds ahead of the excavation equipment and breaks up the existing pavement.
2. *Excavation.* A backhoe shovel excavates and clears the trench for the pipe.

3. *Shoring.* A prefabricated trench "box" made of high-strength steel is placed in the excavated trench to protect the workers in case of a collapse.

4. *Haul.* Debris and broken pavement is removed from the site by trucks.

5. *Manual Labor.* One crew performs final grading of the bed, aids in placing pipe, and installs a rubber gasket around the pipe joint. A second crew does hand backfill and compaction and then forms and pours the concrete paving.

6. *Backfill.* Aside from a small amount done by the laborers, the backfilling operation is done by the backhoe shovel.

7. *Compaction and Preparation for Paving.* This is done by a small roller and manual labor.

8. *Concrete Repaving of the Street.*

The pipe sections are assumed to be large-diameter concrete units (72 in. diameter × 8 ft length) requiring a crane for transport and placement. It is further assumed that the work is pursued in 40-ft sections and that pipe sections and that pipe sections are moved as required from a stockpile next to the route of the excavation. A schematic diagram of this work site is shown in Figure 13.1.

The resources to be considered in this modeling are

1. Pipe units
2. Crane
3. Backhoe shovel
4. Trucks
5. Pavement breaker
6. Trench box
7. Labor crews

An integrated model of the pipelaying process is shown in Figure 13.2. The structure of the model is dictated by the cycle of the mainline sections (40 ft) initiated at QUEUE node 2.

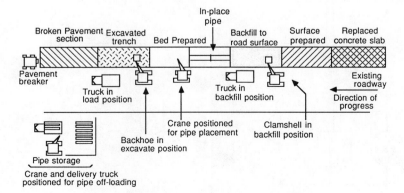

Figure 13.1 Sewer-line job layout.

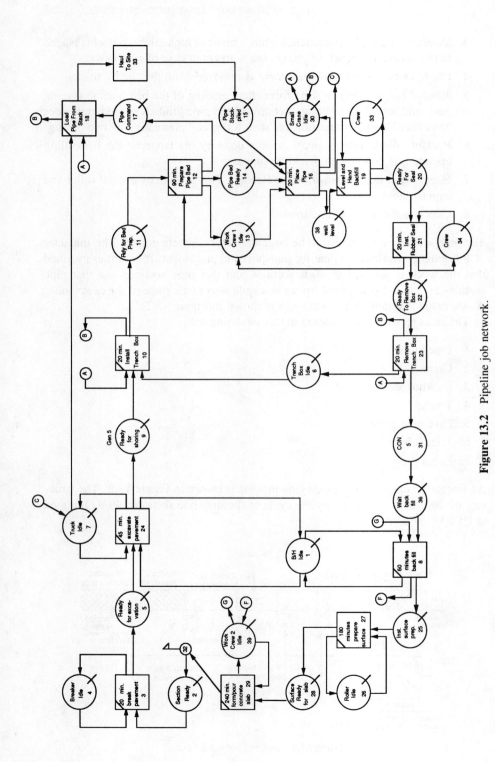

Figure 13.2 Pipeline job network.

238

The section is initially available for pavement removal. This is accomplished at COMBI 3, "Break Pavement." Following this, the pavement is broken and removed and the trench section is excavated using the backhoe shovel. This requires the truck. Once the 40-ft section has been excavated, space is created for the placement of five pipe units. This results in the generation of five units at QUEUE node 9. In the portion of the model between QUEUE node 9 and the FUNCTION node 31, the processed unit is a pipe unit rather than the original 40-ft section of line.

It is assumed that the trench box used to shore the trench is large enough to accommodate one 8-ft length of concrete pipe. For this reason, the GENERATE function is at QUEUE node 9. Had the trench box been 40 ft long, the GENERATE function would be placed at QUEUE node 11. The trench box is placed in the excavation and moved along the line as needed to provide safe working areas for the labor crew. The crane at QUEUE node 30 is used for moving the box.

The individual pipe lengths are hauled from the stockpile once the bed has been prepared at COMBI 12. This is implemented by a message from 12 to the QUEUE node 17 indicating the need for a pipe element. The crane is required for loading the pipe section onto the truck, off-loading the truck, installing the trench box, and removing the trench box.

Following the placement of the pipe, it is leveled and hand-backfilled. CREW 1 supports the preparation of bed and the placement and leveling of the pipe. The CREW 1 resource pattern is a butterfly. This indicates that the crew may give priority to bed preparation over the placement and leveling of the pipe (QUEUE node 12 over 16). Unless more than one "trench box" is defined, CREW 1 will always work the element 12–16–19–21 in sequence. The definition of only one flow unit at QUEUE node 6 ensures that only one pipe bed and unit will be processed at a time.

Once the trench box has been installed and removed 5 times, this indicates that five units of pipe have been installed and the section is ready for backfilling. This is implemented by the CONSOLIDATE function at FUNCTION node 31. The section is then backfilled, prepared for pavement and paved. The completion of the section is registered at COUNTER 32.

Typical time durations for the work tasks are shown in Figure 13.2. The initial positions of the flow units are as follows:

1. Two sections at 2
2. One pavement breaker at 4
3. One truck at 7
4. One backhoe shovel at 1
5. One trench box at 6
6. One crane at 30
7. CREW 1 at 13
8. CREW 2 at 39
9. One roller at 26
10. Crews at 33 and 34

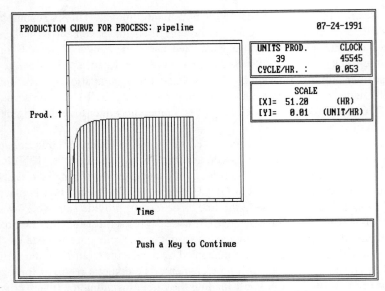

Figure 13.3 Production curve for pipeline job.

The production curve for this process is shown in Figure 13.3. The percent idleness values for the QUEUE nodes in the model are given in Figure 13.4.

13.3 A PRECASTING PROCESS

A cyclic process related to the construction of the rapid-transit rail system in Atlanta (Metropolitan Atlanta Rapid Transit Authority) required the off-site casting of deck sections used in construction of the elevated structures along the line. The process involved the construction of an elevated deck using a precast deck slab for the spans between cast-in-place piers. To accomplish the prefabrication, the precaster designed a concrete stress frame with a raised steel bottom shaped to the contour of the slabs. The raised steel bottom created a plenum into which steam could be fed during the curing. The frame was approximately 20 ft (6.1 m) × 180 ft (54.9 m) and accommodated five slabs at one casting.

The prefabrication operation begins with the forming crew cleaning and oiling the steel forms. Next, the steel side forms are bolted to the steel bed and the wooden end forms are secured to the steel forms. Wood is used on the ends to expedite the stripping of the form. With steel end forms, concrete tends to leak out around prestress strands and delays removal. The wood forms are used once, splintered off, and discarded. A chemical retardant is applied to specific locations of the steel bed where the precast slab, when erected, will eventually make contact with the steel box girder. Since the retardant "kills" the concrete in these areas, later, in the finishing phase, they will be wire-brushed to provide a rough contact surface.

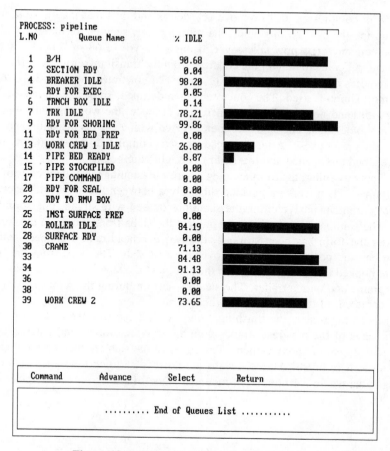

Figure 13.4 Idleness statistics for pipeline job.

With the forms set, the overhead crane drops the pretied bottom layer of steel into position. Then posttensioning ducts are laid in place. The workers now begin pulling 44 strands of $\frac{1}{2}$-in. (12.7 mm)-diameter (seven wire strands) from one end of the stress frame to the other, passing through the wooden end forms. Each wire strand is then cut and secured on both ends of the stress frame.

Safety regulations prohibit any worker from being inside the stress frame during the prestressing process, so the forming crew take their lunch break at this time. During the lunch break, the quality control technician and two workers roll the hydraulic jack into position at one end of the frame and tension each strand to 30,250 lb (13,721 kg). The tensioning takes approximately one minute per strand, and the process is completed in 42 min.

Next, the steel blockouts and inserts are positioned in the form, and the top layer of reinforcing steel is then tied in place. Because of the numerous blockouts and miscellaneous inserts, the top layer of the steel is not pretied into a mat, but, instead, each bar is tied individually in place. As a final step, the forms are blown clean with compressed air to remove any debris and dust that may have accumulated during the day. At this point, the forming crew secures for the day.

The concrete crew now takes over, using a 3.5-yd^3 (2.68 -m^3) bucket and the overhead rail crane to deliver the concrete from the transit mixer to the form. Each form requires approximately 10 yd^3 (7.65 m^3) of concrete, and a $2\frac{1}{2}$–3-in. (63.5–76.2-mm) slump is used. The concrete is then dumped, spread, vibrated, and finished with hand floats. After the concrete is in place, the workers start removing the metal blockouts. The slabs are then covered with a canvas tarpaulin and commence the preset phase. After the preset phase is completed, the steam lines under the steel bed are opened and the 12-hr overnight cure begins.

The next morning, the forming crew begins unfastening the anchor bolts of steel side frames. Then workers position themselves between the slabs with acetylene torches, simultaneously cutting the same prestressed wire strand, one strand at a time. The simultaneous cutting is to prevent the slabs from sliding in the form.

After the form has been stripped, the two overhead cranes connect with the lifting bar and position themselves over the first slab. The lines from the lifting bar are hooked to the lifting loops in the slab, and the slab is then lifted out of the stress frame and onto a trailer. The steam is left on during the removal, since the slabs come out of the forms easier when warm.

Once on the trailer, the finishing crew pries off the wooden end forms, trims off the ends of the prestress strands with an acetylene torch, and patches up the ends with a coat of epoxy cement. The edges of the slab are then dressed up with an electric wire brush. After the slabs are finished, a truck tows each slab to the storage area where a 40-ton truck-mounted crane lifts it into a storage position.

13.4 PRECAST PROCESS ANALYSIS

The CYCLONE model for the precasting process is shown in Figure 13.5. The model was formulated based on a time study that was conducted at the precast plant. The main flow unit is a bulk slab 10 ft by 150 ft (3.05 m by 45.8 m), which in reality consists of the five individual 10 ft by 30 ft (3.05 m by 9.15 m) deck slabs cast in the stress frame. The GENERATE function at node 33 divides the large slab into the five actual slabs.

The production curve and queue idleness information for this model are given in Figures 13.6 and 13.7.

The following information can be developed from the system reports generated by a simulation analysis. Statistics at FUNCTION COUNTER 36 indicate that the first slab was completed after 20.7 working hours. The fifth slab (subelement of the bulk slab) is clocked as completed at time 1290 min or 21.5 working hours of simulated time. A comparison of the simulated results, versus actual field pro-

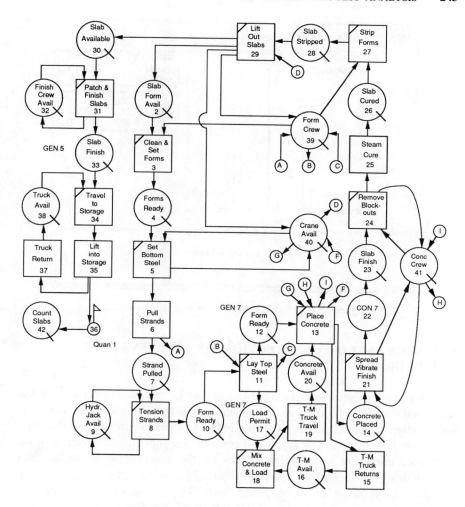

Figure 13.5 CYCLONE precast model.

duction timings, yields the results shown in Table 13.1. The variation between actual and simulated results averages 6% and supports the conclusion that the model is a reasonable replication of the actual process. The system production predicted by the model was also consistent with observed production. The "process report" indicates a production of an hourly production of 0.278 deck sections per production hour.

Once the model is operational and results have been validated as previously indicated, system response to changes in task durations, crane availability, changes in the rate of supplying concrete, and variation in crew size can be studied.

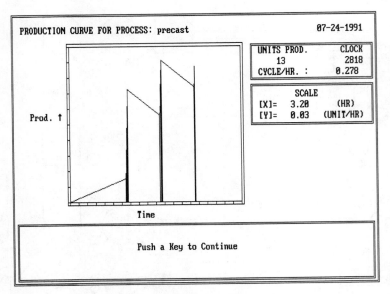

Figure 13.6 Production curve for precasting process.

13.5 SEGMENTED CONSTRUCTION OF AN ELEVATED STRUCTURE

Certain elevated sections of the rapid-transit line in Atlanta were constructed as segmental concrete structures using a technology developed in Europe. This technology might be considered an extension of the deck casting process just described in the previous section. The segmental construction approach has been most notably used elsewhere in the United States for the construction of the Sunshine Skyway Bridge. This was the first application of this modularized concept to constructing a mass transit line in the United States. The process was highly repetitive requiring the casting and erection of 921 integrated box girder and deck sections weighing up to 32 tons each.

Concrete piers on seventy foot centers were constructed along the centerline of the rail line. Concrete segments averaging 10 ft in length and 30 ft in width were fabricated using a "match" casting technique at a construction yard located next to the section of line to be constructed. Once the segments are cast and cured, they are stockpiled on the site and eventually transported by truck to the erection site as required.

The individual segments are lifted onto a temporary truss that spans the 70-ft distance between piers. When all segments of a span are in place, posttensioning cables are run through the deviation blocks of the segments and tied off at the ends in specially designed pier segments. The cable is tensioned and the temporary support truss is moved to the next 70-ft span. The erected segments remain in place in much the same manner that a stack of bricks under compression can be

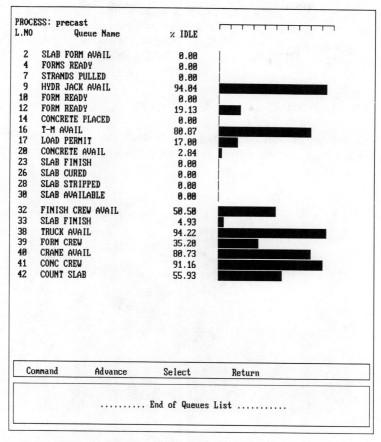

Figure 13.7 QUEUE node idleness statistics for precasting process.

TABLE 13.1 Comparison of Simulated Results and Actual Field Production Timings

(1)	Simulated Completion in Minutes (2)	Actual Time in Minutes (3)
S_1	1242	1145
S_2	1254	1171
S_3	1266	1193
S_4	1278	1211
S_5	1290	1237

held horizontally above the ground provided enough force is applied at the ends of the stack. The erection process is illustrated in Figures 13.8*a–c*.

The matching–casting process will be modeled in the next section. The required activities involved in the fabrication are described below.

13.6 THE MATCHING–CASTING PROCESS

Initially precut and prebent reinforcing steel is stockpiled at the casting yard. A crew of seven steel workers then begins tying the reinforcing steel inside a spe-

(*a*)

(*b*)

Figure 13.8 The segmental construction erection process: (*a*) temporary erection truss in place and ready segments; (*b*) segments hoisted at erection site; (*c*) placement of segments into erection truss.

(c)

Figure 13.8 (*Continued*)

cially designed frame that holds the pieces in place as the cage is being assembled. Pretensioning steel which will give added strength to the flanges of the finished segment is put into place near the top of the cage and tied off at the ends of the frame. Once the cage has been assembled, it is ready to be placed into the oil-treated steel forms.

Before assembly of the forms around the cage can commence, the forms must first be treated with form oil to reduce adhesion between the forms and concrete. The casting bed is also coated with paraffin. Since sections are match–cast to allow for a snug fit between sections during the erection process, the end of the previously cast section is treated to ease separation of the segments when the section now being prepared for casting is cured and ready for stockpiling.

The forms are assembled on the casting bed against the previously cast segment. During the assembly process, the reinforcing steel cage is dropped into place using a special frame (See Fig. 13.9). With the forms, cage, and bulkhead in place, the pouring process may begin.

Concrete [6500 psi (pounds per square inch; $lb/in.^2$)] is brought to the fabrication site as required. Concrete is transferred from the transit mix truck into a 1.5-yd bucket that is hoisted to the top of the forms, where the concrete is dumped into place (see Figure 13.10). A crew of eight workers pours and works the concrete into the forms using hand-held vibrators. After several loads (usually about 17 yd), the form is full and the concrete finishing crew of four workers finishes the exposed surface of the concrete. Also, while the forms are being filled with concrete, test cylinders are made of the concrete placed in the segment.

Specially designed tents are placed over the sections for the overnight steam curing process. Steam produced on the site with a small gas fired steam plant is pumped into the tents throughout the night. The temperature is maintained above

Figure 13.9 Form ready for steel cage.

110°F, while the relative humidity is kept at 100%. In the morning, some 15 hr later, the steam tents are removed and the test cylinders are taken to the testing shed. At this time, the forms are removed but the rebar cage frame remains in place until it is verified that the concrete has reached at least 4000 psi (based on the test cylinders). As soon as this specified strength requirement is met, the ends of the tensioning steel may be cut and the frame removed (see Fig. 13.11).

The segment remains on the casting bed to allow for matching–casting of the next segment. As soon as casting is completed on several consecutive segments, then one can be jacked apart and stockpiled. As the need arises at the construction site, the segments are transported from the stockpile to the erection site.

Figure 13.10 Pouring the concrete.

Figure 13.11 Tensioning steel cut and forms removed.

The end segments, called "pier" segments, are formed in essentially the same way. They, however, require more steel reinforcement and use a different frame and forms. The tying of the pier cage takes approximately 8 hr. The pier segment forms take about 2 hr to remove, versus the typical form removal time of 45 min.

The deterministic times for the model work tasks are given in Table 13.2.

TABLE 13.2 Deterministic Times for Matching–Casting Process

Segment	Time (hr:min)
Typical Segment	
1. Fabricate typical section steel cage	2:30
2. Clean and prepare typical forms	1:00
3. Move typical cage into forms	0:30
4. Set form around cage	1:15
5. Pour concrete	0:45
6. Finish concrete	0:25
7. Cover segment with steam tent	0:20
8. Steam cure (overnight)	15:00
9. Strip forms	0:45
Pier Segment	
1. Fabricate typical section steel cage	8:00
2. Clean and prepare typical forms	1:30
3. Move pier cage into forms	1:30
4. Set forms around cage	2:30
5. Pour concrete	0:45
6. Finish section	0:25

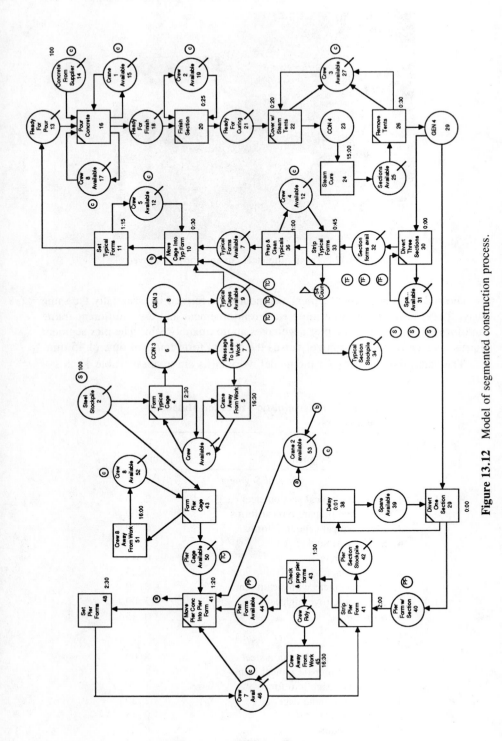

Figure 13.12 Model of segmented construction process.

The CYCLONE model for the segmental construction process is shown in Figure 13.12. Simulation of this model is left as an exercise for the reader.

13.7 CONCLUSION

Although the models presented in Chapters 11 through 13 are necessarily simplified to aid in understanding, they are indicative of the scope of modeling and analysis of construction operations that can be conducted using the CYCLONE system. Use of computers to analyze operations at the job site level has generally been considered too expensive to be justified. Because of the lack of a comprehensive method of investigating complicated construction processes, the attitude has been to rely on "experience" and "engineering intuition." The cost and speculative return on investment of using computerized methods at this level has resulted in little or no analysis being conducted on projects that are not at the superproject level.

The availability of low cost desktop computers has drastically changed the cost effectiveness situation. This availability has virtually eliminated the high risk associated with costly computer procurement. This, in turn, has made methods that rely primarily on computer speed to justify their effectiveness accessible for use in analyzing daily and weekly construction operations. As a result, techniques such as CYCLONE can be readily implemented on a cost-effective basis. In the near future, the use of microcomputers at the job site for the study and analysis of process related problems will be commonplace. The microcomputer in the job trailer will be just as necessary to control the job as the project bar chart.

14 Sensitivity Analysis

14.1 THE MANAGER'S PERSPECTIVE

Managers are continually attempting to evaluate the impact of their decisions on the existing state of the production system and its productive output. They are interested in knowing what will happen if the level or setting of one of the controllable management variables is changed. In short, the sensitivity of the system to management decisions is of the utmost importance.

By experimenting with the available system "controls," a manager can determine how they affect the system and whether they constitute a "brake" or an "accelerator" to system performance. Furthermore, the magnitude of the braking or acceleration can be determined. The ability to go "inside" the system and examine the system parameters and statistics may lead to new insights and indicate the need for a system redesign.

14.2 SYSTEM SENSITIVITY

In order to see how the sensitivity of a system can be investigated to determine the correct levels for management-controlled parameters, consider a problem involving the operation of a precast concrete panel facility. A laboratory experiment on this paper-and-pencil model will be carried out to establish the system's response to parameter variation. This approach is commonly referred to as "sensitivity" analysis.

The CYCLONE model of the plant's process line operation is presented in Figure 14.1. The flow units defined in the system, together with their initial location and flow paths, are given in Table 14.1. The ingredience-constrained activities and their associated ingredience sets are defined in Table 14.2. The number of ingredients required at the ingredience-constrained activities range from four types of units at activity 2 to two types at activities 6 and 19. The unit category types are

1. Space—set and cure positions
2. Materials—forms and batches
3. Equipment—crane and trucks
4. Labor—crews

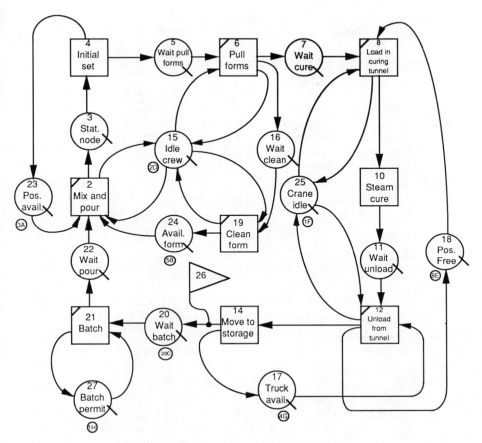

Figure 14.1 Precast concrete element plant.

TABLE 14.1 Entity Path Table

Category	Name	Number Initialized	Initialized At	Path
A	Set positions	3	23	23–2–3–4
B	Forms	5	24	24–2–3–4–5–6–16–19
C	Batches	20	20	20–21–22–2–3–4–5–6–7–8–10–11–12–14–26
D	Crews	2	15	15–2/15–6/15–19
E	Cure position	8	18	18–8–10–11–12
F	Crane	1	25	25–8/25–12
G	Trucks	4	17	17–12–14
H	Batch permit	1	27	21–27

TABLE 14.2   Ingredience-Constrained Work Tasks

			Act		
	2	6	8	12	19
Ingre-dient	Mix and Pour	Pull Forms	Load in Curing Tunnel	Unload from Tunnel	Clean Form
1	Crew at 15	Batch and form at 5	Batch at 7	Batch at 11	Crew at 15
2	Batch at 22	Crew at 15	Cure position at 18	Truck at 17	Form at 16
3	Set position at 23		Crane at 25	Crane at 25	
4	Form at 24				

The crews at QUEUE node 15 and the crane at QUEUE node 25 are shared units. The crew works alternately on activities 2, 6, and 19. The crane works between activities 8 and 12. The "Set" and "Cure Position" units act as simple permits releasing activities 2 and 8, respectively, as appropriate. Once all set positions are committed, no mix-and-pour activity can commence until a position becomes free. Similarly, a casting (batch) cannot be lifted into the curing tunnel until a cure position is available at 18. The system productivity is measured by the counter element at 26. The QUANTITY parameter is set to 10.0 and indicates that 10 elements are associated with each production flow unit. This can be thought of as 10 elements that are simultaneously batched and poured using a single form. The progress of elements through the system from batch to storage is a function of several constraining factors.

The cyclic flow of the forms controls the frequency of the pour–mix operation by virtue of its presence in the associated ingredience set. However, the flow of forms in the system is itself a function of the availability and task selection of the crew units stationed at 15. The labeling sequence is such that the crew units give priority to mixing and pouring, pulling forms, and cleaning forms, in that order. The crane unit at 25 controls flow through the curing tunnel activity and gives priority to loading units over unloading. Truck units at 17 and the availability of the crane control the removal of concrete elements from the curing tunnel. Once units have passed the counter element, they are reentered into the system at a rate defined by the time parameters associated with batch work task 21. Twenty batch units are used initially. This does not result in system constraint for the smaller two- and three-crew systems. However, for systems using four and five crews and six to eight forms, system production constraints will result because 20 batches are used. That is, the mix-and-pour work tasks are at times constrained by the requirement to wait for batches. Therefore, the number of batch concrete element

TABLE 14.3 System Time Attributes

Element Number	Distribu- tion	Set Number	Mean	Lower Bound	Upper Bound	Standard Deviation
2	Normal	2	30.0	20.0	40.0	5.0
4	Constant	3	50.0	0	0	0
6	Constant	4	20.0	0	0	0
8	Normal	5	15.0	8.0	22.0	5.0
10	Constant	6	120.0	0	0	0
12	Normal	5	15.0	8.0	22.0	5.0
14	Normal	7	25.0	10.0	40.0	7.5
19	Normal	8	45.0	17.0	83.0	20.0
21	Normal	9	18.0	9.0	27.0	6.0

units allowable in the system must be increased. Time data for the activities of the process are as shown in Table 14.3.

14.3 PRECAST PLANT SYSTEM RESPONSE

Data reflecting system response are given in Tables 14.4–14.8. These data are based on 100-cycle simulations of systems in which the number of crews and forms available has been varied through a range of values. The number of crews available was varied from two to five, while the number of forms varies from two to eight. These values are considered for the purposes of this illustration to be manager-controlled variables, while other values, such as the number of set positions, were assumed to be fixed and not subject to the control of the manager.

This tabular presentation gives only a small subset of the data generated by the functions associated with this system. However, the tables provide sufficient information to evaluate system response and draw certain conclusions about its performance. The system values monitored are

1. System productivity in units per hour (Table 14.4)
2. Crew availability time (Table 14.5) based on element 15

TABLE 14.4 System Output (Elements per Hour)

Number of Crews	Number of Forms			
	2	4	6	8
2	7.98	12.01	12.16	12.15
3	7.98	15.07	17.02	17.06
4	7.98	15.50	18.02	18.56
5	7.98	15.50	17.88	18.01

TABLE 14.5   Crew Availability Time (%)

Number of Crews	Number of Forms			
	2	4	6	8
2	0.685 / 54%	0.032 / 3%	0.020 / 2%	0.020 / 2%
3	1.680 / 100%	0.549 / 14%	0.118 / 9%	0.077 / 5%
4	2.680 / 100%	1.460 / 100%	0.723 / 51%	0.682 / 48%
5	3.678 / 100%	2.450 / 100%	1.552 / 78%	1.592 / 80%

Time-Integrated Average of Crews 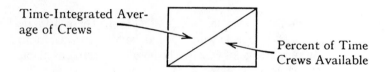 Percent of Time Crews Available

3. Form cycle delays (Table 14.6) to include
   a. Delays because of form pulling (5)
   b. Delays for form cleaning(16)
   c. Delays awaiting other units at the mix-and-pour work task (24)
4. Curing delays information (Table 14.7) to include
   a. Delays awaiting entry to the curing tunnel (7)
   b. Percent of time cure positions are free (18)
5. Crane idle time (Table 14.8 based on 25)

Normal distributions have been associated with all transit time activities with the exception of those having constant durations. Times are developed from field data on observed transit times. The selection of the distribution is also based on examination of the distribution of field transit times.

Examination of the performance data tables indicates certain general information regarding the operation of the system. It will be noted that with the exception of the crew idleness values (Table 14.5), the performance statistics in column one of all tables do not vary as a function of the number of crews. The production of each of these systems is 7.98 units per hour, and the form cycle delays are all zero. This happens because the system is totally constrained by the low number of forms being used and is insensitive to the variation of other control parameters.

TABLE 14.6 Form Cycle Delays

Number of Crews	Number of Forms			
	2	4	6	8
2	0%	36%	52%	69%
	0%	34%	97%	97%
	0%	24%	36%	39%
3	0%	5%	40%	62%
	0%	4%	43%	45%
	0%	11%	50%	86%
4	0%	0%	2%	10%
	0%	0%	4%	13%
	0%	9%	72%	100%
5	0%	0%	0%	1%
	0%	0%	0%	2%
	0%*	0%	75%	100%

*No delay of forms prior to batching.

Pull Delay (5)
Clean Delay (16)
Delay at Mix
and Pour (24)

TABLE 14.7 Curing Delay Information: Percent Time Delay and Time-Integrated Average

Number of Crews	Number of Forms			
	2	4	6	8
2	6%/0.063	5%/0.051	8%/0.08	9%/0.09
	100%/5.92	100%/4.85	100%/4.79	99%/4.73
3	6%/0.063	16%/0.162	22%/0.234	26%/0.30
	100%/5.92	100%/3.79	100%/2.96	94%/2.86
4	6%/0.063	19%/0.214	75%/3.14	65%/2.73
	100%/5.92	100%/3.65	21%/0.71	32%/0.87
5	6%/0.063	19%/0.214	83%/4.89	87%/4.67
	100%/5.92	100%/3.65	16%/0.59	11%/0.53

Wait Cure (7)
Cure Position
Free (18)

TABLE 14.8 Crane Idle Time

Number of Crews	Number of Forms			
	2	4	6	8
2	62%	38%	37%	37%
3	62%	22%	11%	15%
4	62%	21%	5%	6%
5	62%	21%	5%	5%

Crew availability, of course, increases to 100% as the number of crews increases, since two crews are adequate and the additional crews above this value add no productive output.

Another aspect of general system performance that is not indicated in the tables, but is given in the system statistics, is the delay from the time of system "startup" until the first production concrete unit arrives at the counter element (26). This is normally referred to as the "pipeline" delay. For all systems examined, the "pipeline" delays ranged from 270 to 290 min. This means that >4 hr are required for plant startup.

System production increases as a nonlinear function of both the number of crews and forms. Figure 14.2 gives a graphical representation of the information present in Table 14.4. The production for a system using two crews becomes constrained at values of four or more forms, since at this value and above the crews are totally occupied (i.e., saturated). This is verified by the values in the crew availability timetable that shows availability time as 2–3% from four forms and above. The maximum production of this system peaks at 12 elements per hour. Similarly, the three-crew system becomes constrained by the number of

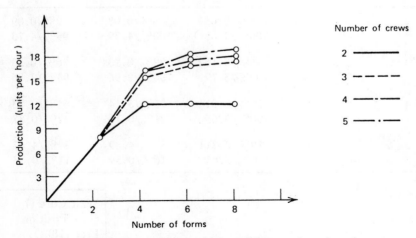

Figure 14.2 Production versus number of forms and number of crews.

crews when six or more forms are available with a peak output of 17 units per hour. The four- and five-crew systems begin to be constrained at values of 18.56 and 18.01, respectively. At this point, the system is limited by the availability of the crane in the curing sequence. The high output of the forming cycle begins to saturate the crane with work, passing production control to the curing segment of the system. The waiting times of the curing facility (element 7) increase significantly from 22% for the three-crew system to 75% for the four-crew system. The average number of curing positions drops from 2.96 to 0.71.

The crew idle time data indicates that the two- and three-crew systems become saturated at form availability values of 6 and 8, respectively. Additionally, with only two forms available, the crews in the two-crew system are fully committed only 46% of the time. The time-integrated average of units idle indicate that 0.68 and 1.680 units on the average are idle in the two- and three-crew systems. Figure 14.3 gives a graphical presentation of the percent time available figures contained in Table 14.5. Although the three-crew system drops from 9 to 5% availability between the six- and eight-form values, the associated increase in production is negligible.

The values given in Table 14.6 indicate the percent of the time concrete units are delayed at the activities in the form cycle (i.e., form pulling, cleaning, awaiting mix, and pour). The delays for pulling and cleaning are directly related to the number of forms and inversely related to the number of crews. This is indicated in the two plots shown in Figure 14.4. Figure 14.4*a* shows the increase in delay

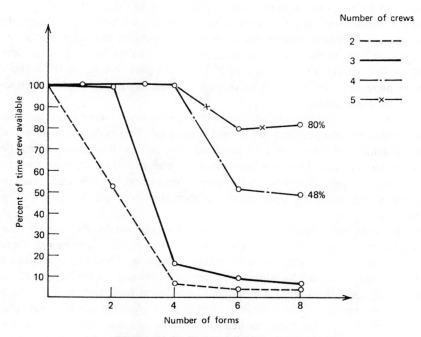

Figure 14.3 Crew availability (percent).

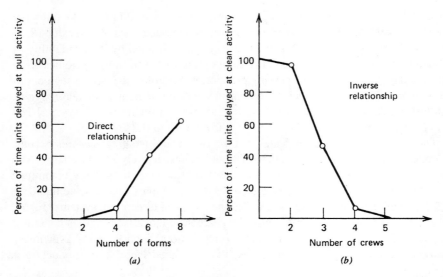

Figure 14.4 Form cycle delay relationships: (*a*) constant number of crews (3); (*b*) constant number of forms (8).

of units at the pull form operation (work task 6) for the three-crew system. This delay increases as the number of forms defined in the system increases. Similarly, Figure 14.4*b* shows that for a constant number of forms in the system, the delay at the clean form operation (work task 19) decreases as the number of crews is increased. In the two- and three-crew systems, delays are large, since the eight forms tend to distribute themselves at these lower-priority activities. Since the crews give priority to the mix-and-pour activity (based on the labeling rule), these "back-cycle" activities must wait longer for a crew assignment. As the number of crews increases, the crew/form ratio improves, causing a corresponding reduction in form cycle delays.

Form cycle delays at the mix-and-pour activity (because of other units) increase as the number of forms defined in the system increase. This is due to the effect of the fixed number of three initial set positions that, in this illustration, cannot be varied by the manager. This constraint places an upper limit on the rate at which elements can be generated from the pouring–forming cycle. This effect can be observed by examining the delay values in each row from left to right. This is illustrated in Figure 14.5.

Delays at the curing activity segment are not significant for the smaller two- and three-crew systems. The larger systems (e.g., five crews, eight forms), these delays become important and constrain system production. The delay is a function of both availability of cure locations in the curing tunnel and the availability of the crane to load and unload units. For the five-crew, eight-form system, Table 14.7 shows that the percent of the time cure positions are available awaiting new elements is only 11%. This indicates that all positions are committed 89% of the time. Furthermore, the crane is committed 95% of the time (5% idle time), as

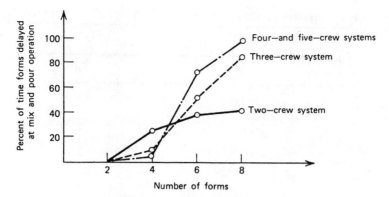

Figure 14.5 Mix and pour delays.

indicated in Table 14.8. As noted in the discussion of system productivity, the saturation of the crane also begins to constrain the system and causes the peak system output to be governed by the curing network segment instead of by the form cycle, as in the smaller systems. The decrease in curing location availability and the increase in element delay time at the curing tunnel loading operation are most pronounced in the 4–6, 4–8, 5–6, and 5–8 systems.* This is also the case with the crane availability, which is illustrated in Figure 14.6. As the crane availability is reduced, the system approaches a deterministic limit of 20 units/hr. The crane makes an average of four loads–unloads per hour based on the 15-min mean load and unload duration associated with these activities. This results in an average of two loads and two unloads per hour. The two unloading work tasks generate 20 elements for storage.

*The first value indicates the number of crews and the second the number of forms.

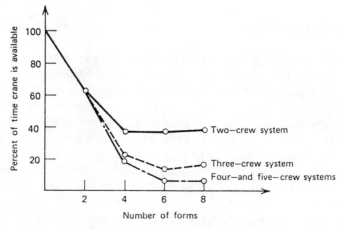

Figure 14.6 Crane availability.

TABLE 14.9   Interarrival Rate at Curing Tunnel in Minutes (Element 7)

Number of Crews	Number of Forms		
	4	6	8
3	38.11	33.11	33.36
4	37.15	28.90 (32.95)	28.50 (32.13)
5	37.15	27.90 (33.26)	27.60 (33.12)

In the section on productivity, it was noted that higher system productivities were achieved with the four-crew systems (18.02 and 18.56) than with the five-crew systems (17.88 and 18.01). This reduction in system productivity can be traced to congestion in the curing tunnel segment of the model. This congestion is increased by the higher arrival rate of elements at the curing tunnel from the forming cycle. Table 14.9 shows the interarrival rates for the nine largest systems investigated. The increase in crews and forms leads to a higher production of elements from the forming cycle and a corresponding higher interarrival rate at the curing tunnel. Table 14.7 indicates that the percent of the time curing locations are available in the four-crew system is considerably greater than in the five-crew systems. This results in a faster throughput and higher overall productivity. This is indicated by the interarrival rates at work task 14 following the curing tunnel (values in Table 14.9 in parentheses). This indicates, for instance, a greater interarrival rate at 14 for the 5–8 system than for the 4–8. Obviously, delays in the larger system have occurred in the curing segment.

14.4   COST ANALYSIS

The analysis conducted in this example has focused only on the plant output in terms of units-per-hour production. A deeper analysis must also address the cost per unit, which requires additional input features designed to associate cost attributes with productive units (crews, crane, etc.) as appropriate. For instance, if it is assumed that the hourly cost associated with each crew is $200, the cost range involved in the experiments conducted above varies from $400 to $1000 (assuming other costs to be fixed). Table 14.10 shows a cost-per-element variation based on these assumed crew costs. The cost analysis indicates that the two-crew system yields the lowest cost per unit ($3.26) with six or more forms. However, if demand is strong, the best combination of production and element cost probably occurs in the three-crew, six-form system, which produces 17 elements per hour at a cost of $3.52 per element.

TABLE 14.10 Cost-per-Element Variation

Number of Crews	Number of Forms			
	2	4	6	8
2	5.02	3.33	3.26	3.26
3	7.54	3.98	3.52	3.52
4	10.06	5.18	4.43	4.31
5	12.58	6.45	5.60	5.53

The object of this presentation has not been to examine the precast plant system exhaustively but, instead, to indicate the insights into system operation that can be developed using simulation and the modeling simplicity afforded by the CY-CLONE system format. The interaction of the various flow units and the process activities becomes considerably clearer after an investigative simulation of the system such as the one conducted here on the concrete plant problem. If the manager desires a more detailed analysis, this can be carried out using more sophisticated statistical analysis methods. If, however, the manager simply wants to get the "feel" of the system and the interactive forces and constraints at work, the simple qualitative analysis carried out in this example can be used. The depth of analysis is obviously a function of the detail and statistical reliability of the information desired and the resources available. In most construction systems, the precision of the input data (e.g., transit times, distribution types) is based on the engineering estimates and a sampling of field data. This precision seldom justifies the high degree of statistical confidence appropriate in other applications. It does, however, provide sufficient accuracy to carry out an insight-oriented analysis such as the one just developed. In most cases, this type of analysis is most appropriate for construction management purposes.

14.5 HIGH-RISE CASE STUDY

Now let us consider the results of the simulation of the construction of the Peachtree Plaza Tower model originally presented in Chapter 12 (Sections 12.4–12.8). Data regarding floor cycle times for various combinations of crews and shear wall forms were given in Table 12.2. The model for this system is presented here as Figure 14.7 for reference.

Utilization of the cranes for the cases presented in Table 12.2 can be developed from the idle time figures given in Table 14.11. Variation of the number of the CREW 2-type units from one to two has negligible effect on system response. The 2×1 system configuration yields the most economic utilization values for the cranes, resulting in the large crane (at 2) being active on system tasks 38% of the time and the smaller crane being active 32% of the time for the 10 shear wall case. This activity is confined to tasks defined in the concrete placement model, and the

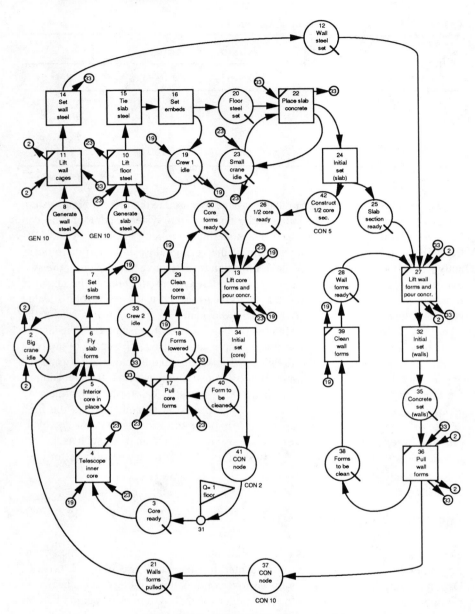

Figure 14.7 CYCLONE process model.

cranes will be available for other tasks 62% and 68% of the time, respectively. Actual observed idle times at the job site are reduced by loading the cranes with lifting tasks not defined in the concreting model (e.g., lifting glazing and plumbing fixtures, interior finish items). The 2×2 (two type 2 crews and two type 1 crews) configuration results in only a slight increase in the utilization of the large crane, while the small crane utilization remains the same.

TABLE 14.11 Crane Idle Times

Number of CREW 1	Number of CREW 2	Ten Shear Wall Forms	Eight Shear Wall Forms	Five Shear Wall Forms	Two Shear Wall Forms
1	1	69 / 77	69 / 77	70 / 78	
2	1	62 / 68	68 / 71	68 / 75	78 / 78
1	2	68 / 77	68 / 77	69 / 77	
2	2	58 / 68	59 / 68	64 / 70	72 / 73

(In each cell: Percent Time Idle (Large Crane) / Percent Time Idle (Small Crane))

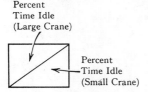

Percent Time Idle (Large Crane)

Percent Time Idle (Small Crane)

Crew utilization on systems tasks can be developed from Table 14.12. The occupancy statistics for the 1 × 1 system configuration indicate that the CREW 1 unit is highly committed, being active on system tasks 88% of the time. The CREW 2 unit is busy only 48% of the time in this configuration. Increasing the number of CREW 1 units, with the attendant increased input to the shear wall concreting sequence, causes an increase in the activity of the CREW 2 unit to 62% of the time. The CREW 1 occupancy statistic rises sharply, since there are now two units.

The other tables presented (see Tables 14.13–14.16) indicate the delays incurred by units at various critical points in the flow network. The effect of the increase input rate from the slab steel placement sequence (10–15–16) caused by increasing CREW 1-type units can be observed by considering the idle times at QUEUE nodes 28 and 25 as given in Table 14.15. In the 1 × 1 configuration, units at 25 are delayed only 6% of the time in the 10- and 8-form systems. This is increased to 17% in the 2 × 1 system, since the addition of an extra CREW 1 unit causes higher productivity along the 10–15–16 track and results in a greater backup of units at this monitoring point. The 1 × 2 system records no delays at 25, since the single CREW 1 units now become the constraining unit, causing the release units from 12 to be delayed. The constraining track in every case can be determined by examining the preceding QUEUE node delays and evaluating the idle times. In the 2 × 2 with only eight shear wall forms, the occupancy statistic

TABLE 14.12 Crew Idle Times

Number of CREW 1	Number of CREW 2	Ten Shear Wall Forms	Eight Shear Wall Forms	Five Shear Wall Forms	Two Shear Wall Forms
1	1	12 / 52	12 / 52	14 / 54	
2	1	60 / 38	57 / 44	62 / 48	73 / 62
1	2	12 / 85	12 / 85	13 /	
2	2	65 / 87	60 / 88	59 / 87	65 / 86

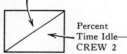

Percent
Time Idle—
CREW 1

Percent
Time Idle—
CREW 2

TABLE 14.13 Delay at 6

Number of CREW 1	Number of CREW 2	Ten Shear Wall Forms	Eight Shear Wall Forms	Five Shear Wall Forms	Two Shear Wall Forms
1	1	0 / 6	0 / 6	0 / 7	
2	1	3 / 1	9 / 1	16 / 1	31 / 1
1	2	0 / 6	0 / 5	0 / 7	
2	2	0 / 3	2 / 1	2 / 1	29 / 1

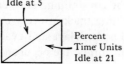

Percent
Time Units
Idle at 5

Percent
Time Units
Idle at 21

TABLE 14.14 Delay at 4 and 22

Number of CREW 1 \ Number of CREW 2		Ten Shear Wall Forms	Eight Shear Wall Forms	Five Shear Wall Forms	Two Shear Wall Forms
1	1	3 / 23	3 / 23	3 / 21	
2	1	1 / 29	1 / 29	1 / 25	1 / 19
1	2	3 / 8	3 / 8	3 / 8	
2	2	2 / 15	1 / 15	1 / 15	1

Percent Time Units Idle at 3
Percent Time Units Idle at 20

TABLE 14.15 Delay at 27

Number of CREW 1 \ Number of CREW 2		Ten Shear Wall Forms	Eight Shear Wall Forms	Five Shear Wall Forms	Two Shear Wall Forms
1	1	6 / 100	6 / 100	22 / 73	
2	1	17 / 96	17 / 90	25 / 73	46
1	2	0 / 99	0 / 99	6 / 76	
2	2	6 / 94	13 / 83	24 / 70	51 / 34

Percent Time Units Idle at 25
Percent Time Units Idle at 28

TABLE 14.16 Delay at 13

Number of CREW 1	Number of CREW 2	Ten Shear Wall Forms	Eight Shear Wall Forms	Five Shear Wall Forms	Two Shear Wall Forms
1	1	4 / 94	4 / 94	7 / 94	
1	2	0 / 82	0 / 81	0 / 84	0 / 84
2	1	4 / 95	4 / 95	6 / 99	
2	2	0 / 82	0 / 81	0 / 88	0 / 82

Percent
Time Units
Idle at 26

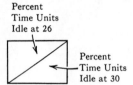

Percent
Time Units
Idle at 30

that previously indicated a very high availability of clean forms at 28 drops sharply, reflecting for the first time potentially significant delays at 27 caused by lack of clean shear wall forms.

The delays at element 6 occur primarily as a result of (1) the requirement to wait until all 10 wall forms have been pulled at 36 or (2) the completion of telescoping of the interior core forms. The delays are indicated by the idle times associated with QUEUE nodes 5 and 21, as given in Table 14.13. In the system with one unit of each crew type, the delay of zero associated with QUEUE nodes establishes that the permit at 21 is always present when required and that commencement of form movement (flying) is never delayed because of the pulling of shear wall forms. When the number of CREW 1-type units is increased to two, the situation reverses with delays usually occurring because of the requirement to wait until all forms are pulled. That is, the idle time at 5 now usually exceeds that at 21. This develops because the additional crew at 19 expedites flow of the core sections along segment 13–34 as well as at the interior form telescope operation. This effect becomes more pronounced as the number of forms is reduced.

The idle times in Table 14.16 indicate that in the systems utilizing two CREW 1-type units, the core pour is never delayed because of either nonavailability of the forms, the crane, or the crew required. The 1 × 1 and 1 × 2 system do show some delay at 26 because of the requirement to compete for the single CREW 1 unit with the slab steel placement sequence (10–15–16).

15 Noncyclic Networks

All the networks previously discussed in this text have been cyclic as the name CYCLONE implies. That is, the networks are closed cycles and resources remain within the networks. It is possible, however, to use CYCLONE to model non-cyclic networks. In this situation, primary resource entities traverse the network only once and then exit. Examples of noncyclic networks are schedule networks and decision trees. These types of noncyclic networks are covered in this chapter.

15.1 SCHEDULE NETWORKS

One type of noncyclic network is the schedule network. Network scheduling techniques are used extensively to determine a project's duration. Two of the most popular techniques are the Critical Path Method (CPM) and the Probabilistic Evaluation and Review Technique (PERT). This section describes a method to simulate schedule networks using CYCLONE. It is assumed most readers have some familiarity with CPM and PERT. Therefore, the following discussion centers around the limitations of CPM and PERT and how computer simulation might overcome some of these limitations.

In spite of the widespread acceptance of CPM and PERT, schedule overruns continue to be a major problem. One possible reason is that network schedules calculated with CPM or PERT do not provide adequate information regarding the potential for schedule overruns. That is, CPM gives only a single number, which is intended to be the duration of a project. PERT is but a slight improvement in that it attempts to evaluate the probability of project duration by giving the expected completion time. Additionally, the PERT method sums the variance of the activities along the path used to calculate the expected completion time in order to express a measure of risk of the project duration.

Although PERT introduces elements of probability into the calculations, PERT consistently underestimates a project's duration. The principal cause of this underestimation is a condition known as ''merge event bias.'' Briefly, merge event bias occurs when several paths converge on a single node. Figure 15.1 is a simplified depiction of how several paths in a schedule network might converge on a single node.

PERT calculations give the early expected finish time of this node as the summation of times on the longest path leading to the node. This path then becomes part of the longest path through the network that determines expected project duration. However, since the duration of the activities on the paths are random var-

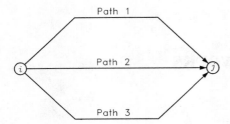

Figure 15.1   Merge event bias.

iables, it is possible that some other path converging on the node could have an activity with a random duration longer than its expected (mean) duration. Thus, this longer path would determine the early finish time of the node. That this potential longer path is not taken into account in the PERT calculation leads to an underestimation of project duration.

Additionally, the PERT method assumes statistical independence between activities. This assumption allows the variance of activities along a path to be added, giving the variance of the duration of the project. The assumption of independence, however, may not always be appropriate. For instance, weather can create a positive correlation between activities, and a delay in one activity may create a negative correlation between activities.

One solution to the difficulties noted above is computer simulation. Because Monte Carlo simulation of schedule networks does not use a single number to represent activity durations, it avoids the merge event bias described above.

15.2   CYCLONE–CPM NETWORK

The conversion of a CPM network to a CYCLONE network is relatively straightforward. CPM activities become either NORMAL or COMBI activities in CYCLONE. Figure 15.2 illustrates the correspondence between CPM and CYCLONE. In Figure 15.2a, a CPM activity path between nodes i, j, and k becomes a CYCLONE series of NORMAL work tasks. In this instance, the CYCLONE work task is roughly equivalent to an activity in a precedence network. There is no constraint on activity B other than the completion of activity A in either CPM or CYCLONE. In Figure 15.2b, activity D cannot start until the completion of activities A, B, and C. In the CYCLONE network, this relationship is modeled with a COMBI work task as follows.

Work tasks A, B, and C all release a single entity to their following QUEUE node. Further, entities must be present in all three QUEUE nodes in order for COMBI D to start. The above considerations are essentially all that are required to convert a CPM network in CYCLONE; however, the relationships in Figure 15.2c may be instructive for the somewhat more complex condition shown. Again (except for the QUEUE nodes), the similarity to precedence networks is evident. And, it should be noted, the similarity to PERT networks is also evident.

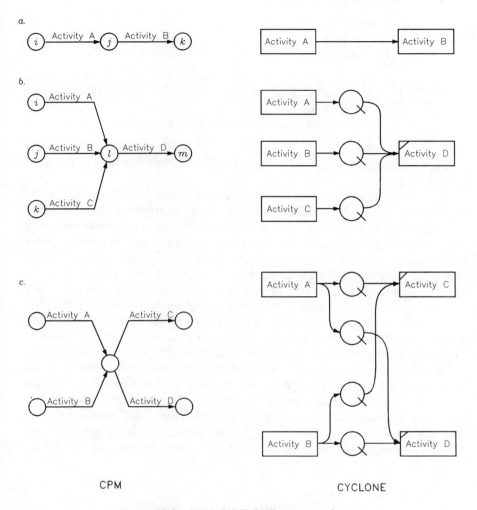

Figure 15.2 CPM–CYCLONE correspondence.

One final observation before discussing an example of network simulation by CYCLONE is that CYCLONE can incorporate probabilistic arcs to model random paths emanating from a NORMAL or COMBI work task. Also CYCLONE can incorporate feedback loops to represent a rework condition, for example.

15.3 EXAMPLE PROJECT

The example project which follows is the overhaul of a hypothetical crude oil heater unit. To introduce the concept, the project is obviously oversimplified and resources are not used. Figure 15.3 is the CPM diagram of the project. Estimated

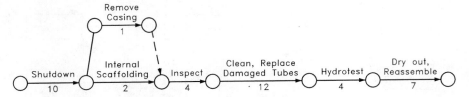

Figure 15.3 Overhaul bolder CPM network.

durations are in days. Figure 15.4 is the CYCLONE representation of the CPM network of Figure 15.3.

By way of illustration, notice the estimated duration for "hydrotest" in the CPM network must take into account the possibility of repairing leaks and retesting. In the CYCLONE network, this possibility can be incorporated more precisely with the probabilistic nodes and feedback loop to the activity "Hydrotest." Now "Hydrotest" can have a more appropriate duration. It should be noted, parenthetically, that PERT and CPM do not have this capability. Elements 4, 5, and 17 serve to introduce a single entity to start the process, and element 16 serves as a counter in order to simulate the network as many times as desired. Other than these elements, the entry of data into a CYCLONE network does not involve any more overhead than entering data into a CPM network. The network was simulated 100 times and the nonconstant activities were modeled by beta (β) distributions.

The output for the overhaul boiler CPM network is shown in Table 15.1. The column marked "COUNT" signifies that the network was simulated 100 times. Note that "hydrotest" was simulated 112 times, reflecting the 12 times the path was taken to repair leaks. The count of 12 for "Repair Leaks" confirms that the path was taken approximately 10% of the time as specified in the network input. Likewise, as expected, the numbers under the column headed "MEAN DUR"

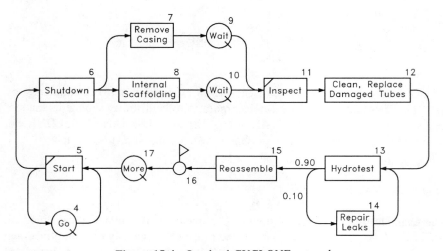

Figure 15.4 Overhaul CYCLONE network.

TABLE 15.1 CYCLONE Output for Boiler Overhaul

TYPE	LABEL	DESCRIPTION	COUNT	MEAN.DUR	AR.TIME	AVE.NUM	%BUSY
COMBI	3	BEGIN	100	0.00	37.15	0.00	0.0
NORMAL	6	SHUT DOWN	100	9.77	37.25	0.26	26.3
NORMAL	7	REMOVE CASING	100	1.00	37.26	0.03	2.7
NORMAL	8	INTERNAL SCAFF	100	2.00	37.27	0.05	5.3
COMBI	11	INSPECT	100	4.00	37.31	0.10	10.4
NORMAL	12	CLN REPL TUBES	100	12.08	37.42	0.32	32.1
NORMAL	13	HYDROTEST	112	2.00	33.48	0.06	6.0
NORMAL	14	REPAIR LEAKS	12	3.85	312.28	0.01	1.2
NORMAL	15	REASSEMBLE	100	7.01	37.57	0.19	18.7

TYPE	LABEL	DESCRIPTION	AVG.WAIT	AVG.UNIT	UNITS END		%OCCUPIED
QUE	2	READY	37.20	1.0	1		100.0
QUE	9	CASING REM'D	0.95	0.0	0		2.6
QUE	10	SCAFF SET	0.00	0.0	0		0.0
QUE	17	MORE	0.00	0.0	1		0.0

TYPE	LABEL	DESCRIPTION	COUNT	BETWEEN	FIRST
FUN	16		100	37.57	37.06

denote a mean duration for the activities very close to the numbers shown for the activity duration input in Figure 15.3. The remaining statistics, "AR.TIME," "AVE.NUM," and "% BUSY" are not especially meaningful in the context of this simulation.

The meaningful statistic is at the bottom of Table 15.1 listed as the function node number 16. Here, the number 37.57 under the heading "BETWEEN" is the average number of days for project completion. This statistic is not affected by the PERT network merge node bias discussed earlier in this article. It is acknowledged, however, in this simple network, merge node bias would not be a factor. The project duration is shorter than that which would be calculated by PERT or CPM because the loop for "Repair Leaks" permits an alternative modeling of the "Hydrotest" activity. In other words, the duration of 4 days for "Hydrotest" in the CPM network included the possibility of repairing leaks, whereas in the CY-CLONE network, "Hydrotest" and "Repair Leaks" can be separated and treated probabilistically.

The average number of days for project completion is obviously an important statistic since it indicates that 50% of the time the expected time of completion will be less than 37.57 days and 50% of the time the expected time of completion will be greater than 37.57 days.

A somewhat troublesome aspect in the interpretation of any computer simulation of a network regards the critical path. Computer simulation packages do not typically calculate a critical path since it could change with succeeding passes through the network. In CYCLONE, one way to identify likely critical paths would be to examine the statistics under "AVG.WAIT" in Table 15.2. This statistic indicates the amount of time an entity had to wait in a QUEUE node before being processed by the following COMBI activity. QUEUES with the shortest percent time occupied will be the most likely to lie on the critical or longest path through the network. To illustrate in the simple demonstration network, it is clear the critical path would not include activity 7, "Remove Casing." This is reflected in the percent occupied time of 2.6 of QUEUE node 9 compared to the other QUEUE nodes with 0.0% time occupied.

15.4 DECISION TREE NETWORK

Decision trees are graphical methods for analyzing alternative choices and the outcomes that result from making a choice. Outcomes in a decision tree are determined by chance. Consider the decision tree of Figure 15.5.

Here, a contractor has a choice of bidding on project A or B. If the contractor chooses to bid on project A, a choice is to be made whether to assign a 0%, 1%, or 2% markup. Further, if the contractor assigns a markup of 0%, it is assumed the probabilities of winning and losing are 0.9 and 0.1, respectively.

Should the contractor win project A with a 0% markup, the profit for the project is described by three chance branches with a probability of 0.3, 0.4, and 0.3. Profits associated with these probabilities are $10,000, $15,000, and $17,000. The

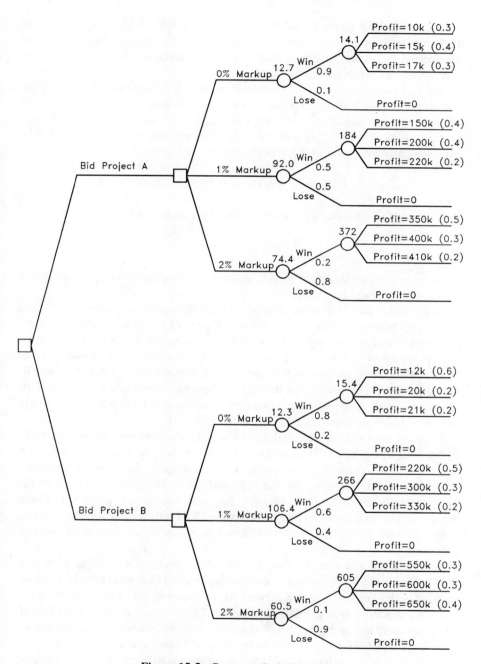

Figure 15.5   Contractor's decision tree.

other choices for projects A and B are as shown in Figure 15.5. Note in Figure 15.5 that decision nodes are usually represented as squares and chance nodes are represented as circles. Also note that the decisions and outcomes are portrayed in chronologic sequence from left to right.

The traditional method of solving decision trees is to work backward from the end nodes using a form of expected value. The two most common forms of expected value are expected monetary value and expected utility value. In this example, expected monetary value is used. The method works as follows. The expected monetary value of the uppermost node in Figure 15.5 is

$$0.3(10) + 0.4(15) + 0.3(17) = 14.1k$$

This calculation is simply the expected value of a probability distribution:

$$E(X) = \sum p_i x_i \qquad (15.1)$$

where p_i is the probability of a realization of the random variable X and x_i is a corresponding realization of X.

In a similar way, the other expected monetary values are calculated. These values are shown above the choice nodes of Figure 15.5. Now the branches of the tree can be pruned. For project A, the highest expected monetary value ($92,000) is associated with a 1% markup and so the other branches are eliminated. Likewise, a 1% markup for project B is optimal and between project A and project B, project B with a 1% markup has the highest expected monetary value ($106,400). Therefore, a contractor wishing to maximize the expected monetary value would choose to bid project B with a 1% markup.

Decision trees can also be analyzed by simulation. The tree can be structured in the form of a CYCLONE model and run as a noncyclic network. Figure 15.6 is a CYCLONE network of the decision tree of Figure 15.5. Choice nodes are depicted as function nodes, and the end branches of the decision tree are NORMAL tasks. In the CYCLONE network, however, the discrete probability distributions of the decision tree have been replaced by continuous triangular distributions. Also the NORMAL work task durations are in dollar units rather than in time units.

Note that the win–lose combination for each increment of markup ends in a QUEUE node. This is to ensure a resource entity exits from either the win or lose node of each choice before the network is simulated again. Thus, one resource entity is collected at each of QUEUE nodes 23–28 before being released to COMBI 29. Counter 30 tallies the number of times the network has been simulated before releasing to QUEUE node 31, where the process begins again.

Table 15.2 is a summary of selected output for the network. For this example, the network has been simulated 100 times. The relevant statistics from Table 15.2 are the counts and mean durations.

The relationship between the count and mean duration for each of the NORMAL activities of the CYCLONE network and the choice nodes of the tree net-

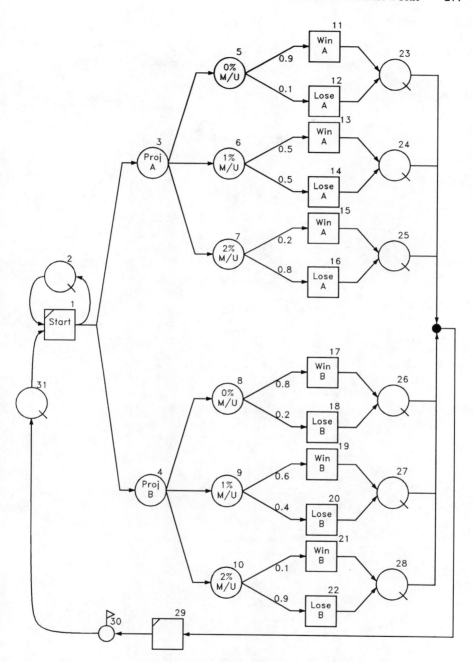

Figure 15.6 CYCLONE network of decision tree.

TABLE 15.2 Partial CYCLONE Output for Decision Tree

TYPE	LABEL	DESCRIPTION	COUNT	MEAN.DUR	AR.TIME	AVE.NUM	%BUSY
NORMAL	11	WIN A @ 0%	89	14.08	284.01	0.05	5.0
NORMAL	12	LOSE A @ 0%	11	0.00	2233.81	0.00	0.0
NORMAL	13	WIN A @ 1%	54	191.38	458.37	0.41	40.9
NORMAL	14	LOSE A @ 1%	46	0.00	549.16	0.00	0.0
NORMAL	15	WIN A @ 2%	19	385.47	1313.73	0.29	29.0
NORMAL	16	LOSE A @ 2%	81	0.00	311.87	0.00	0.0
NORMAL	17	WIN B @ 0%	85	17.78	297.39	0.06	6.0
NORMAL	18	LOSE B @ 0%	15	0.00	1403.68	0.00	0.0
NORMAL	19	WIN B @ 1%	45	284.82	561.36	0.51	50.7
NORMAL	20	LOSE B @ 1%	55	0.00	459.30	0.00	0.0
NORMAL	21	WIN B @ 2%	8	602.42	2997.23	0.19	19.1
NORMAL	22	LOSE B @ 2%	92	0.00	252.78	0.00	0.0

work is as follows. NORMAL activity 11 is related to winning project A with a 0% markup. The mean duration value of 14.08 for NORMAL 11 corresponds to a value of 14.10 for the decision tree of Figure 15.5. Although the values are approximately equal, this proximity should not be expected to hold throughout because there is not a direct correspondence between the discrete probability distributions of the decision tree and the continuous probability distributions of the CYCLONE network.

Continuing the analysis of Table 15.2, the count of 89 for NORMAL 11 indicates that entities passed through this node 89 out of 100 cycles corresponding to the 0.9 probability assigned to the arc leading to NORMAL 11. Multiplying 89/100 times the mean duration of 14.08 gives an expected value of 12.53 or $12,530. Continuing similar calculations for the remaining NORMAL tasks of Table 15.1 produces the maximum expected monetary value of $128,170 for NORMAL 19 (win project B with a 1% markup) confirming the previous analysis of the decision tree.

Simulation may not provide an advantage over traditional decision tree analysis in rather simple examples such as the one shown here. However, analysis of more complicated networks might prove to be more efficient by simulation. Moreover, the use of continuous distributions in simulation could allow for more comprehensive modeling of probability distributions.

15.5 DESIGN NETWORK

In the first example of this chapter, a CPM network was converted into a CYCLONE network. In many instances, the CYCLONE network can be drawn directly as shown in Figure 15.7. This figure is an illustrative CYCLONE network for a design process.

To briefly review the network, COMBI 1 along with QUEUE nodes 2 and 3 introduce a single resource entity into the network to start the simulation. Observe that "Concept Plans" and "Site Plans" both must be complete before "Engr Analysis" can begin. Similar relationships hold for the "Preliminary Cost Est" and all the reviews that follow. COUNTER node 46 keeps track of the number of cycles the network has been simulated and sends a signal back to QUEUE node 3 to start the process over again. In the CYCLONE input for this model the activities are modeled by triangular distributions.

Typically, when analyzing a CYCLONE model of a scheduling network, one is interested in the critical activities, the effect of adding or decreasing resources, and perhaps making probability statements about the duration of the process.

As in the first example of this chapter, potential critical activities may be identified by examining the waiting time in QUEUE nodes. Table 15.3 is the QUEUE node report for the design network. For the three QUEUE nodes preceding COMBI 19, "Prelim Cost Est," QUEUE node 16 had an average wait time of 0.0. This means resource units were waiting in QUEUE nodes 15 and 17 while a resource unit was being processed by NORMAL 12, "Floor Plans," on its way to QUEUE

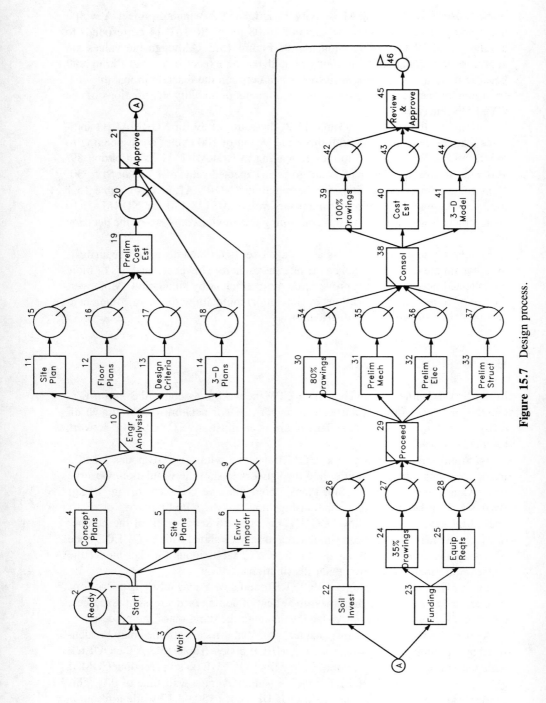

Figure 15.7 Design process.

TABLE 15.3 QUEUE Node Report for Design Network

TYPE	LABEL	DESCRIPTION	AVG.WAIT	AVG.UNIT	UNITS END	%OCCUPIED
QUE	2	READY	240.50	1.0	1	100.0
QUE	3	WAIT	0.00	0.0	0	0.0
QUE	7		2.36	0.0	0	1.0
QUE	8		0.00	0.0	0	0.0
QUE	9		28.03	0.1	0	11.5
QUE	15		10.76	0.0	0	4.4
QUE	16		0.00	0.0	0	0.0
QUE	17		1.35	0.0	0	0.6
QUE	18		10.76	0.0	0	4.4
QUE	20		0.00	0.0	0	0.0
QUE	26		51.59	0.2	0	21.1
QUE	27		0.00	0.0	0	0.0
QUE	28		25.08	0.1	0	10.3
QUE	34		0.00	0.0	0	0.0
QUE	35		31.79	0.1	0	13.1
QUE	36		26.79	0.1	0	11.0
QUE	37		30.84	0.1	0	12.7
QUE	42		0.00	0.0	0	0.0
QUE	43		21.29	0.1	0	8.8
QUE	44		10.84	0.0	0	4.5

TYPE	LABEL	DESCRIPTION	COUNT	STATISTICS BETWEEN	FIRST
FUN	46		100	242.90	249.78

281

node 16. Also note, however, that QUEUE node 17 has a relatively low waiting time—an indication that NORMAL 13, "Design Criteria," is also a critical activity candidate. Therefore the likely critical path is along nodes 5–10–12–19–21–23–24–29–30–38–39–45 with nodes 4 and 13 potentially lying on a parallel critical path.

The technique of varying resources in the network, called *sensitivity analysis*, has been discussed in Chapter 14. In the design network of Figure 15.7, no physical resources are used. Accordingly, one would see the *implied* effect of varying resources by changing the durations of the appropriate NORMAL or COMBI work tasks.

Briefly the basis for probability statements about simulated schedule networks is the following. For any particular run of a network with probabilistic durations, the duration of the network is the sum of a sequence of random variates. This sum is a random sample from simulating the network one time. It follows that the average duration for a given number of runs is the average of a sum of random samples. From the central limit theorem of statistics, the probability distribution of the sum of a number of random samples approaches a normal distribution.

In order to make probability statements about normal distributions, the parameters of the distribution μ and σ must be known or estimated. A common technique for estimating μ and σ is to use the sample mean and standard deviation as point estimates of the mean and standard deviation of the parent population. For a more complete discussion of this technique, see Ang and Tang (1975). Now the probability that a random sample from a normal distribution is less than or equal to a specified number, x, is given as follows:

$$P(X \leq x) = P\left(z \leq \frac{x - \mu}{\sigma}\right) \tag{15.2}$$

The mean of the schedule network, μ, is given in Table 15.3 in the statistics for the FUNCTION node. In this case the average duration is 242.9 days. The standard deviation of 5.15 was taken from runs using the sensitivity analysis module of CYCLONE. Please see the user manual for a description of this module.

The probabilistic interpretation of the average duration is that there is a 50% chance the project will take less than or equal to 242.9 days. One might like to know, for example, the probability that the design project will be completed in 250 days or less. This probability is

$$P(X \leq 250) = P\left(z \leq \frac{250 - 242.90}{5.15}\right) = P(z \leq 1.38)$$

From a table of standard normal probability values, the probability that $z \leq 1.38$ is 0.92 or 92%.

Other probability statements can also be made. For instance, the number of days one can be sure of completing the project with 95% confidence is given by setting the expression for z in equation (15.2) to a value from the standard normal probability table corresponding to the desired percentage.

Again, from probability tables, the value of z corresponding to a probability of 0.95 is 1.65. Setting the expression below for z equal to 1.65 and solving for x gives the number of days as 251.4 or 251:

$$P\left(z \leq \frac{x - 242.90}{5.15}\right) = 0.95$$

The models presented in this chapter are unlike the CYCLONE networks depicted earlier in this book in that they do not follow the familiar cyclic patterns. These noncyclic networks are intended to point out that CYCLONE can be applied to a wide range of problems.

APPENDIX A
Exponential Assumptions and Infinite Queues

A.1 EXPONENTIALLY DISTRIBUTED ARRIVAL AND SERVICE RATES

The assumption of exponentially distributed arrival and service times greatly simplifies the mathematical definition of a queueing system and the determination of the state probabilities $P_i (i = 0, M + 1)$.

The mathematical expectation of an arrival in the interval Δt is

$$\frac{1}{\lambda} = \int_0^\infty tf(t) \, dt$$

For exponentially distributed arrival times

$$f(t) = \lambda e^{-\lambda t} \tag{A.1}$$

The probability that no arrival occurs in the interval $(0, t)$ is the same as the probability that the first arrival occurs after T.

$$P[t_a \geq T] = \int_T^\infty \lambda e^{-\lambda t} \, dt = e^{-\lambda t} \tag{A.2}$$

Therefore

$$P \begin{bmatrix} \text{No arrival} \\ \text{in interval} \\ \Delta t \end{bmatrix} = P[t \geq \Delta t] = e^{-\lambda \Delta t} \tag{A.3}$$

The expression obtained by expanding this expression $(e^{-\lambda \Delta t})$ as a Taylor series is

$$
P \begin{bmatrix} \text{No arrival} \\ \text{in interval} \\ \Delta t \end{bmatrix} = e^{-\lambda \Delta t} = 1 - \lambda\,\Delta t + \frac{(-\lambda\,\Delta t)^2}{2!} + \frac{(-\lambda\,\Delta t)^3}{3!} + \ldots
$$

$$(A.4)$$

If the higher-ordered terms are neglected (i.e., Δt very small) **the expression reduces to**

$$
P \begin{bmatrix} \text{No arrival} \\ \text{in interval} \\ \Delta t \end{bmatrix} = 1 - \lambda\,\Delta t
\qquad (A.5)
$$

The probability of an arrival in Δt is

$$
P \begin{bmatrix} \text{One arrival} \\ \text{in interval} \\ \Delta t \end{bmatrix} = \lambda\,\Delta t
\qquad (A.6)
$$

Since the assumption is made that the server times are exponentially distributed, the probability of no service completing in the interval $(0, T)$ is the probability that the service completes after T:

$$
P[t_3 \geq T] = \int_T^\infty \mu e^{-\mu t}\, dt = e^{-\mu t}
\qquad (A.7)
$$

where $1/\mu$ = the mathematical expectation of a service completion. Using the Taylor series expansion as above, the probability of no service in interval t becomes

$$
P \begin{bmatrix} \text{No service} \\ \text{in interval} \\ \Delta t \end{bmatrix} = 1 - \mu\,\Delta t
\qquad (A.8)
$$

and the probability of one service in t is

$$
P \begin{bmatrix} \text{One service in} \\ \text{interval } \Delta t \end{bmatrix} = \mu\,\Delta t
\qquad (A.9)
$$

A.2 INFINITE POPULATION QUEUEING MODELS

For systems with infinite populations and exponential arrival and server distributions, the transition probability defined in Chapter 3, equation (3.1) becomes

$$T_{nn} = \{(1 - \lambda\,\Delta t) \times (1 - \mu\,\Delta t)\} + \{(\lambda\,\Delta t) \times \mu(\Delta t)\}$$

$$= 1 - (\mu + \lambda)\,\Delta t + \lambda\mu(\Delta t)^2 + \lambda\mu(\Delta t)^2 \qquad \text{(A.10)}$$

Again, neglecting the higher-order terms, since Δt is taken as very small, this reduces to

$$T_{nn} = 1 - (\mu + \lambda)\,\Delta t \qquad \text{(A.11)}$$

The transition probabilities $T_{(n+1)n}$ and $T_{(n-1)n}$ become

$$T_{(n+1)n} = \{(1 - \lambda\,\Delta t) \times (\mu\,\Delta t)\} \cong \mu\,\Delta t \qquad \text{(A.12)}$$

$$T_{(n-1)n} = \{(\lambda\,\Delta t) \times (1 - \mu\,\Delta t)\} \cong \lambda\,\Delta t \qquad \text{(A.13)}$$

Therefore, the probability of being in state n, $P_n(t + \Delta t)$, at $t + \Delta t$ becomes

$$P_n(t + \Delta t) = [1 - (\mu + \lambda)\,\Delta t]P_n + \mu\,\Delta t[P_{n+1}] + \lambda\,\Delta t[P_{n-1}]$$

or

$$P_n(t + \Delta t) = P_n(t) - (\mu + \lambda)P_n(t)\,\Delta t + \mu P_{n+1}(t)\,\Delta t + \lambda P_{n-1}(t)\,\Delta t \quad \text{(A.14)}$$

Transferring $P_n(t)$ to the left side of the equation and dividing by Δt, the expression is

$$\frac{P_n(t + \Delta t) - P_n(t)}{\Delta t} = -(\mu + \lambda)P_n(t) + \mu P_{n+1}(t) + \lambda P_{n-1}(t) \quad \text{(A.15)}$$

Let $\Delta t \rightarrow 0$ (become infinitesimally small); then

$$\frac{dP_n(t)}{dt} = -(\mu + \lambda)P_n(t) + \mu P_{n+1}(t) + \lambda P_{n-1}(t) \qquad \text{(A.16)}$$

If the system is assumed to be operating under steady-state conditions, then the probability of being in state n (i.e., P_n) does not change with time and

$$\frac{dP_n(t)}{dt} = 0$$

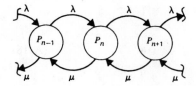

Figure A.1 State S_n in Markov chain.

Therefore,

$$-(\mu + \lambda)P_n(t) + \mu P_{n+1}(t) + \lambda P_{n-1}(t) = 0 \tag{A.17}$$

Considering this situation as a Markovian chain, the model appears as shown in Figure A.1. The state probabilities have been associated with the "lily pads" in this diagram, and the frog leaps are set equal to μ for leaps going up one state and λ for leaps carrying the system down one state. Then equation A.17 can be written directly from the diagram by summing the flows into the node representing state n and setting them equal to the flows out. Flows into S_n are contingent on being in the state from which the flow originates (e.g., S_{n-1} or S_{n+1}). Therefore, they are multiplied by the state probability of the state from which they originate. Flows out of S_n are contingent on being in S_n, and therefore are multiplied by P_n. Therefore, the balance becomes

$$\text{Flow out} = \text{flow in}$$

$$\mu P_n + \lambda P_n = \mu P_{n+1} + \lambda P_{n-1}$$

or

$$-(\mu + \lambda)P_n + \mu P_{n+1} + \lambda P_{n-1} = 0$$

Using this concept, the diagram of state 0, S_0, is given by Figure A.2. The equation for P_1 in terms of P_0 is

$$\text{Flow out} = \text{flow in}$$

$$\lambda P_0 = \mu P_1$$

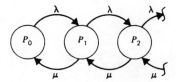

Figure A.2 State S_0 in Markov chain.

or

$$P_1 = \frac{\lambda}{\mu} P_0 \qquad (A.18)$$

This is also written $P_1 = \rho P_0$, where $\rho = \lambda/\mu$ is called the *utilization factor*. The general solution of the set of equations

$$\mu P_1 - \lambda P_0 = 0$$

and

$$\mu P_{n+1} + \lambda P_{n-1} - (\lambda + \mu) P_n = 0 \qquad (A.19)$$

defining a queueing model with infinite population and exponentially distributed arrival and service times when solved in terms of P_0 becomes

$$P_n = \left(\frac{\lambda}{\mu}\right)^n P_0 \qquad (A.20)$$

Since $1/\lambda = T_{\text{arrival}} =$ the mathematically expected or average arrival time, and $1/\mu = T_{\text{service}} =$ the mathematically expected or average service time, then

$$P_n = \left(\frac{\lambda}{\mu}\right)^n P_0 = \left[\frac{T_{\text{service}}}{T_{\text{arrival}}}\right]^n P_0$$

If, for instance, the average arrival time is 20 min and the average service time is 5 min, then

$$P_n = \left(\frac{5}{20}\right)^n P_0 = \left(\frac{1}{4}\right)^n P_0$$

It must be kept in mind that the equations and solution developed here are for a single-channel infinite population system with exponentially distributed arrival and server times. It should also be kept in mind that the solution was obtained by setting $dP/dT = 0$ and therefore applies only to systems that are operating in a steady state. Solution of the equations for transient conditions is more mathematically complex and requires definition of the number of units in the system at time zero.

APPENDIX B
Queueing Nomographs

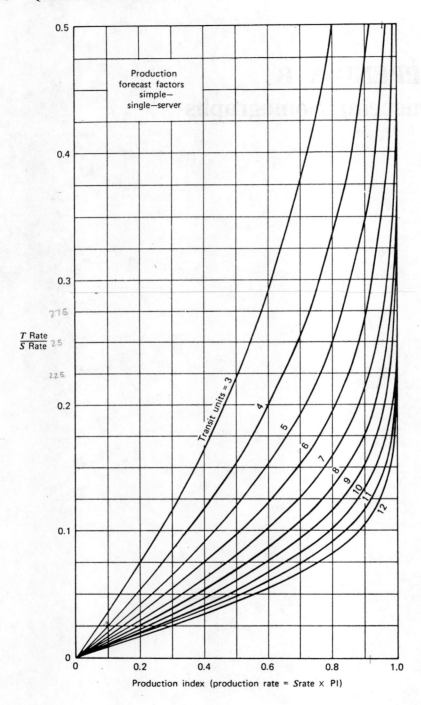

Production forecast factors simple—single—server

$\dfrac{T \text{ Rate}}{S \text{ Rate}}$

Transit units = 3

4

5

6

7

8

9

10

11

12

Production index (production rate = Srate × PI)

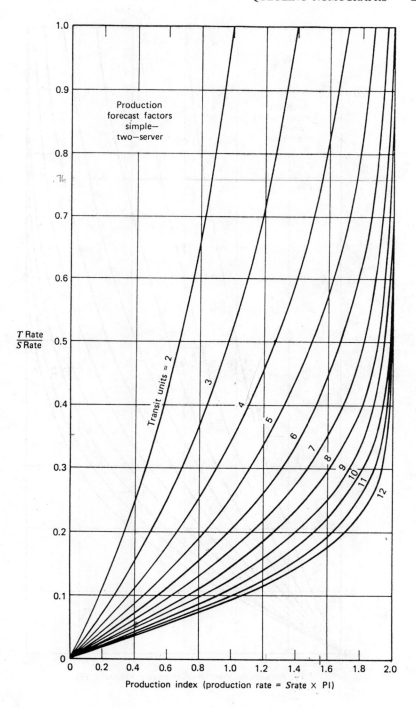

Production forecast factors simple— two—server

$\dfrac{T \text{ Rate}}{S \text{ Rate}}$

Transit units = 2

3

4

5

6

7

8

9

10

11

12

Production index (production rate = Srate × PI)

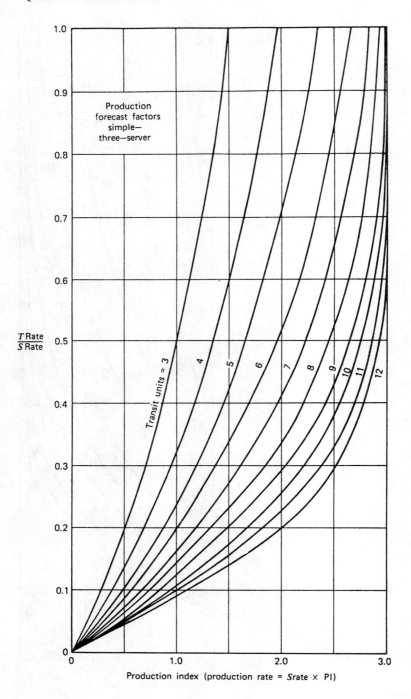

Production
forecast factors
simple—
three—server

$\dfrac{T \text{ Rate}}{S \text{ Rate}}$

Transit units = 3   4   5   6   7   8   9   10   11   12

Production index (production rate = Srate × PI)

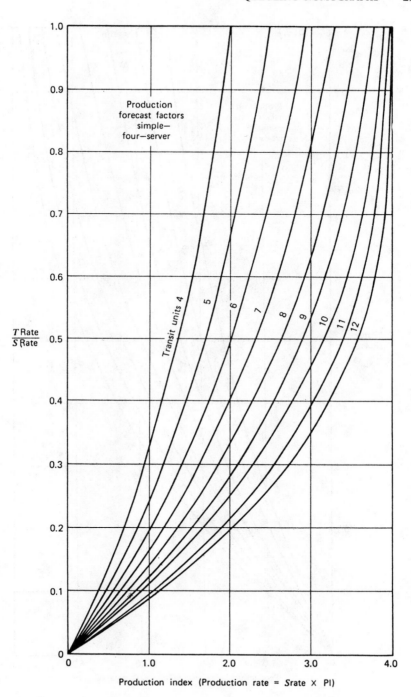

Production index (Production rate = Srate × PI)

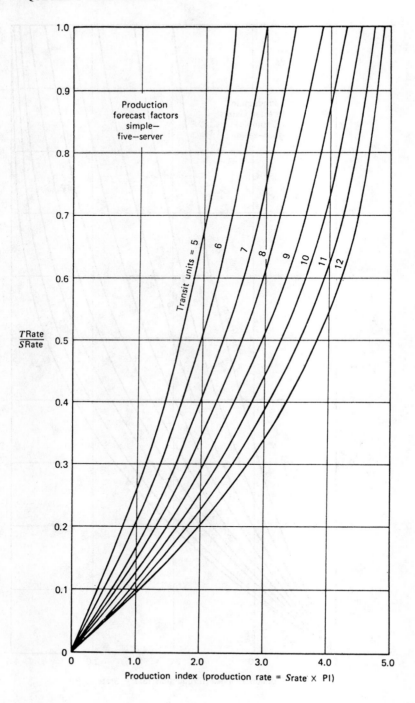

Production
forecast factors
simple—
five—server

Transit units = 5   6   7   8   9   10   11   12

$\dfrac{T\text{Rate}}{S\text{Rate}}$

Production index (production rate = Srate × PI)

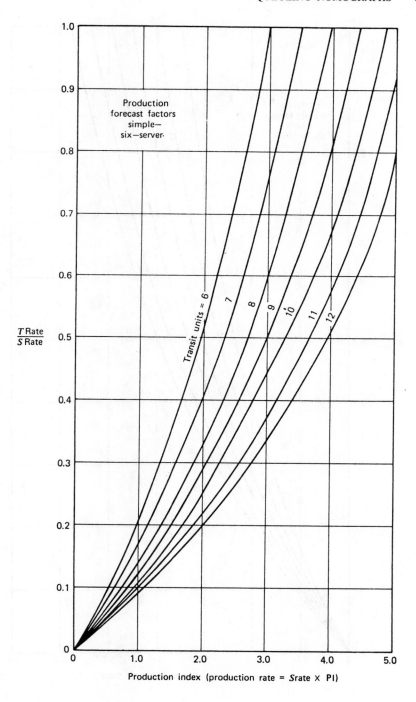

Production index (production rate = Srate × PI)

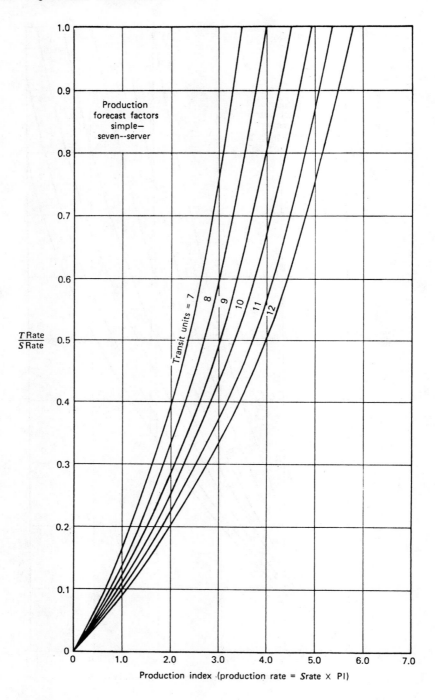

Production
forecast factors
simple—
seven--server

Transit units = 7   8   9   10   11   12

$\dfrac{T \text{ Rate}}{S \text{ Rate}}$

Production index (production rate = Srate × PI)

Hopper—Single—Server
Production forecast factors   Hopper Capacity = 1   Transit units = 2

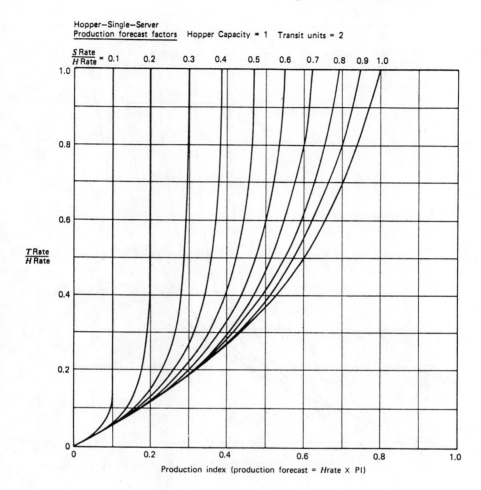

$\frac{S \text{ Rate}}{H \text{ Rate}}$ = 0.1   0.2   0.3   0.4   0.5   0.6   0.7   0.8   0.9   1.0

$\frac{T \text{ Rate}}{H \text{ Rate}}$

Production index (production forecast = Hrate × PI)

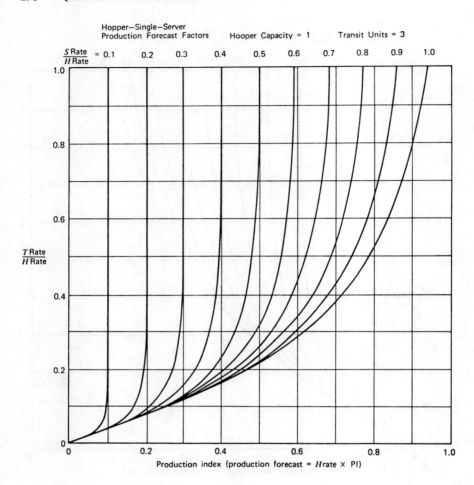

Hopper—Single—Server
Production Forecast Factors   Hooper Capacity = 1   Transit Units = 3

$\frac{S\,\text{Rate}}{H\,\text{Rate}}$ = 0.1   0.2   0.3   0.4   0.5   0.6   0.7   0.8   0.9   1.0

$\frac{T\,\text{Rate}}{H\,\text{Rate}}$

Production index (production forecast = H rate × PI)

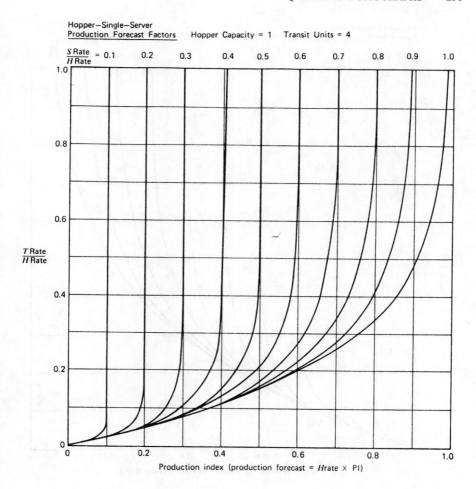

Hopper—Single—Server
Production Forecast Factors   Hopper Capacity = 1   Transit Units = 4

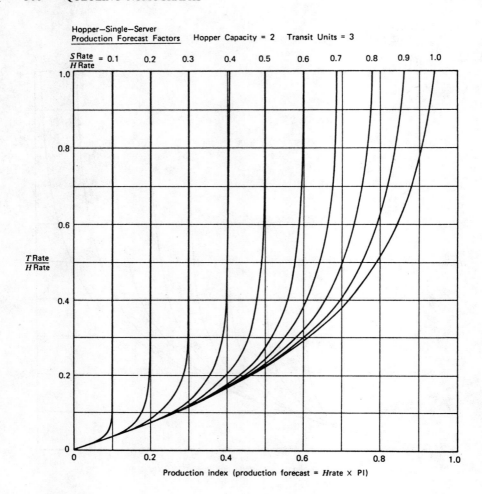

Hopper–Single–Server
Production Forecast Factors   Hopper Capacity = 2   Transit Units = 3

$\frac{S\,\text{Rate}}{H\,\text{Rate}}$ = 0.1   0.2   0.3   0.4   0.5   0.6   0.7   0.8   0.9   1.0

$\frac{T\,\text{Rate}}{H\,\text{Rate}}$

Production index (production forecast = Hrate × PI)

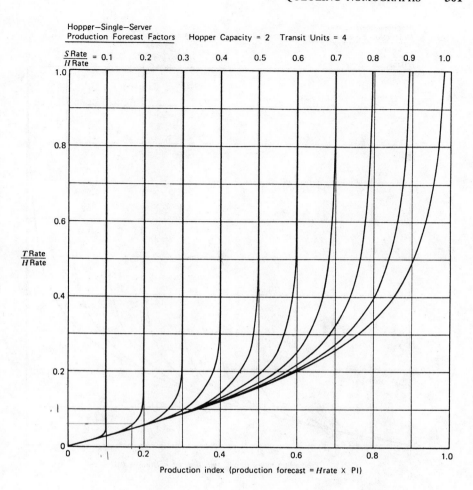

Hopper—Single—Server
Production Forecast Factors   Hopper Capacity = 2   Transit Units = 4

$\dfrac{S\,\text{Rate}}{H\,\text{Rate}}$ = 0.1   0.2   0.3   0.4   0.5   0.6   0.7   0.8   0.9   1.0

$\dfrac{T\,\text{Rate}}{H\,\text{Rate}}$

Production index (production forecast = H rate × PI)

Hopper—Single—Server
Production Forecast Factors   Hopper Capacity = 2   Transit Units = 5

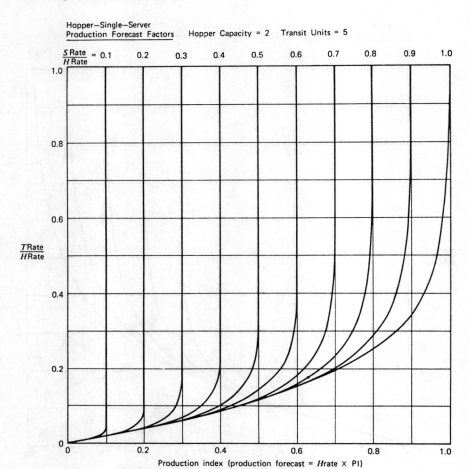

Hopper—Single—Server
Production Forecast Factors   Hopper Capacity = 3   Transit Units = 4

$\frac{S\,\text{Rate}}{H\,\text{Rate}}$ = 0.1   0.2   0.3   0.4   0.5   0.6   0.7   0.8   0.9   1.0

$\frac{T\,\text{Rate}}{H\,\text{Rate}}$

Production index (production forecast = Hrate × PI)

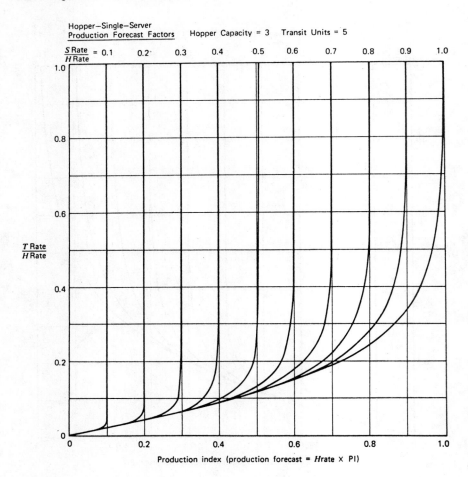

Hopper–Single–Server
Production Forecast Factors   Hopper Capacity = 3   Transit Units = 5

$\dfrac{S\,\text{Rate}}{H\,\text{Rate}}$ = 0.1   0.2   0.3   0.4   0.5   0.6   0.7   0.8   0.9   1.0

$\dfrac{T\,\text{Rate}}{H\,\text{Rate}}$

Production index (production forecast = Hrate × PI)

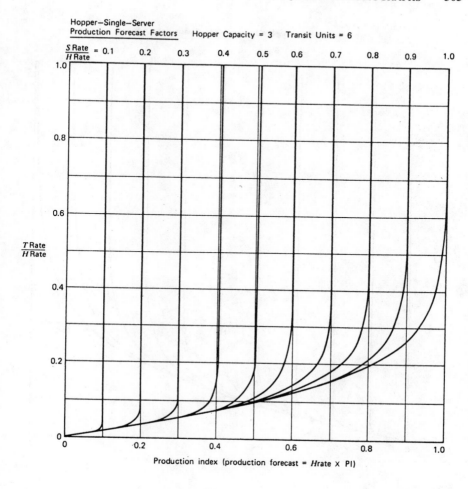

Hopper–Single–Server
Production Forecast Factors   Hopper Capacity = 3   Transit Units = 6

$\frac{S \text{ Rate}}{H \text{ Rate}}$ = 0.1   0.2   0.3   0.4   0.5   0.6   0.7   0.8   0.9   1.0

$\frac{T \text{ Rate}}{H \text{ Rate}}$

Production index (production forecast = Hrate × PI)

Hopper—Single—Server
Production Forecast Factors   Hopper Capacity = 4   Transit Units = 5

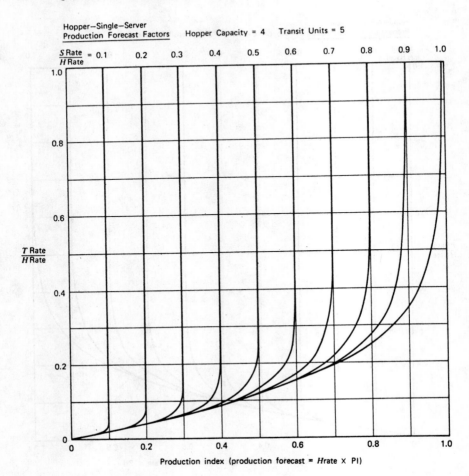

Production index (production forecast = *H*rate × PI)

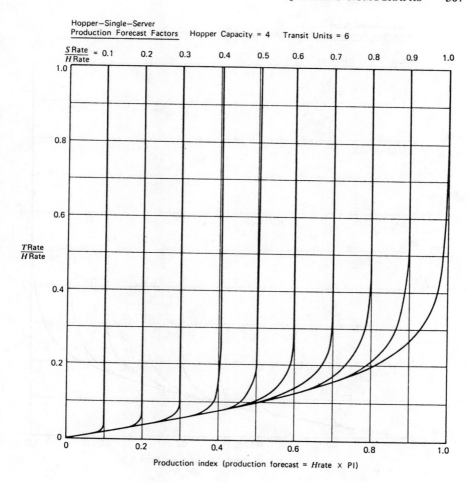

Hopper—Single—Server
Production Forecast Factors   Hopper Capacity = 4   Transit Units = 6

$\frac{S\,\text{Rate}}{H\,\text{Rate}}$ = 0.1   0.2   0.3   0.4   0.5   0.6   0.7   0.8   0.9   1.0

$\frac{T\,\text{Rate}}{H\,\text{Rate}}$

Production index (production forecast = Hrate × PI)

Hopper—Single—Server
Production Forecast Factors   Hopper Capacity = 4   Transit Units = 7

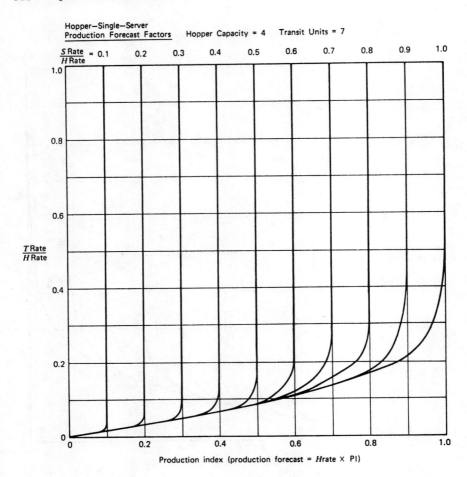

Production index (production forecast = Hrate × PI)

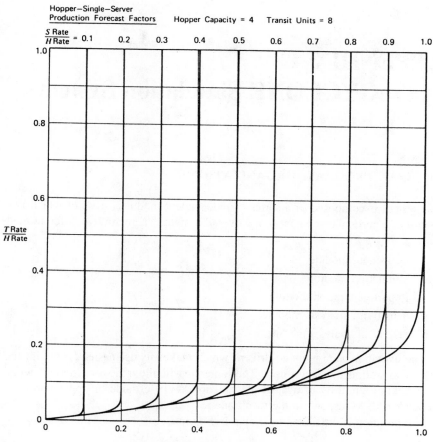

Hopper—Single—Server
Production Forecast Factors   Hopper Capacity = 4   Transit Units = 8

$\frac{S\ \text{Rate}}{H\ \text{Rate}}$ = 0.1   0.2   0.3   0.4   0.5   0.6   0.7   0.8   0.9   1.0

$\frac{T\ \text{Rate}}{H\ \text{Rate}}$

Production index (production forecast = Hrate × PI)

APPENDIX C
MicroCYCLONE Simulation System

GENERAL PROGRAM ORGANIZATION

The system is composed of a series of independent modules, each of which is in control of a particular segment of the overall system. There are five different modules:

1. Data-input module
2. Simulation module
3. Report generation module
4. Sensitivity analysis module
5. Statistical analysis module

The overall system is menu-driven, which makes its use easier for the inexperienced as well as the expert user. These menus will allow the user to move within the whole program by using just the function keys (left side or upper portion of the keyboard) to respond to the menu queries.

The program is organized as shown in Figure C.1.

CONVENTIONAL NETWORK INPUT

Introduction

MicroCYCLONE supports two forms of network input; the standard CYCLONE models and the conventional input form. The following text covers the more widely used conventional input format.

A MicroCYCLONE input file is the means by which the user translates a CYCLONE graphical model into a Problem-Oriented Language (POL) that can be understood by the MicroCYCLONE programs. A graphical CYCLONE model can be fully defined in a POL format by reducing the model into four categories of information as follows:

1. General information about the model, such as its name and the mode by which simulation is terminated.
2. Information about the CYCLONE network, where each node is fully defined with regard to its type, label, and description and its logical relation to other nodes.

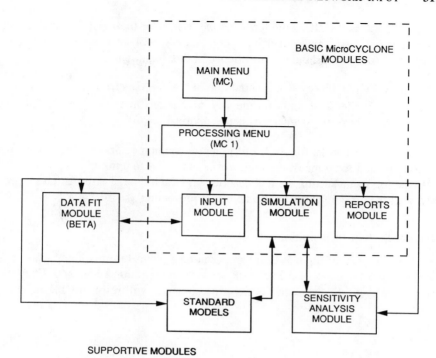

Figure C.1   Organization of MicroCYCLONE package.

3. Information about the duration of work tasks, with regard to the type of distribution used and its parameters.
4. Information about the resources defined in the model, pertaining to their numbers and location at the beginning of simulation.

The specification of the four types of information is structured in an input file format as shown below:

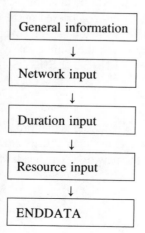

The MicroCYCLONE compiler helps the user enter the input file and edit it, when required, and transform it into compiled code for later use in simulation. Two technical aspects should be mentioned at this stage, namely:

1. All CYCLONE key words should be in CAPITAL letters.
2. The compiler looks only for the *first three* letters in any key word. The rest of the letters are disregarded, hence optional.

The compiler checks for some of the more common errors entered incorrectly into the file and for some of the logic errors but is not foolproof.

In the following section we will describe how each type of information is specified with the proper syntax and some of the procedures.

General System Information

The first category of input is entitled "General System Information." This segment contains the name of the network and certain program parameters. When the user has selected the data entry option, the computer will respond with the command:

LINE # 1 ?

This is the first line of information concerning the network, and it must *always* contain the standard header for general system information, which is defined as follows:

*NAM*E (name of process) *LEN*GTH (length of run) *CYC*LES (# of cycles)

NAME	User-defined keyword assigned to this network.
LENGTH	The length of time, in time units, for the simulated running of the process.
CYCLES	Maximum number of cycles that will be processed during simulation. This will be determined by the number of times the COUNTER function is passed in the process model.
NOTE	The simulation will stop at whichever comes first, the timed length of the run or the number of cycles.
EXAMPLE	NAME TUNNEL LENGTH 100 CYCLES 10

Network Input Segment

This segment of input is used to enter the actual process network. Each statement of this segment specifies one network element, its attributes, and its logical relationship to other elements in the network. The header for this segment is

NETWORK INPUT

The header should be typed in LINE #2.

Four types of elements are used in MicroCYCLONE networks:

1. COMBI
2. NORMAL
3. QUEUE
4. FUNCTION

Each individual element should be entered in a separate line. If the element will not fit on one line of the screen, the user can keep typing on to succeeding lines up to a total of 254 characters. The program will automatically enter this as a single line or record.

COMBI Work Tasks

The following attributes are used to define the COMBI:

- Numeric label
- Element type
- Work task description (optional)
- Duration set number
- Preceding QUEUE nodes
- Following nodes

The general form of the input statement for a COMBI element is

(label.C) COMBI 'descr.' SET (set) PREC (labels.p)
FOLL (labels.F)

where

1. All underlined letters are key words and should appear in the line as shown.
2. All words enclosed in parentheses ( ) would not be entered as typed above, their corresponding entries are described below.
3. (Label.C): is the numeric label (i.e., integer) of COMBI being specified.
4. 'descr.': is a description of the COMBI specified.
5. (Set): is the number (i.e., integer label) of the set where the duration parameters of this COMBI are specified.
6. (Labels.p) are the labels of the preceding nodes of this COMBI.
7. (Labels.F) are labels of the following nodes of this COMBI.

Example:

17 COMBI 'LOAD TRK' SET 6 PRECEDERS 2 5 FOLLOWERS
9 11 15

This specifies a COMBI with the numerical label 17 and the description is 'load trk'. Its corresponding time parameters would be defined in SET 6. It is preceded by the nodes 2 and 5 and succeeded by the nodes 9, 11, and 15.

Note that we would have used the first three letters of any of the key words like FOL for FOLLOWERS.

Normal Work Tasks

Specification of the NORMAL work task is similar to the COMBI except that the preceding operations need not be specified. The attributes required for a NORMAL element are:

- Numeric label
- Element type
- Work task description (optional)
- Duration set number
- Following nodes

The general form of input for a NORMAL is

```
(Label) NORMAL 'descr.' SET (set #) FOLLOWERS
(label of fol.)
```

Example:

```
23 NORMAL 'Trk Return' PARAMETER SET 4 FOLLOWERS
27 30
```

Queue Nodes

The following attributes are required to define a QUEUE node, or a QUEUE node acting as a GENERATE function:

- Numeric label
- Element type
- QUEUE node title (optional)
- GENERATE function and number (when required)

The general format for a QUEUE node is

```
(Label) QUEUE 'description'
```

The general format for a GENERATE function is

```
(Label) QUEUE 'description' GENERATe (number to be
generated)
```

Examples:

5 QUEUE 'Loader Idle'
9 QUEUE 'Truck Queue' GENERATE 5

Function Nodes

Two separate function nodes are used in MicroCYCLONE: COUNTER and CON-SOLIDATE. The general format for the accumulator (counter) function node is:

(Label.C) FUNCTION COUNTER FOLL (Label.F) QUANTITY
(Quant.)

where

1. All underlined letters are key words and should be entered as shown above.
2. Words enclosed in parentheses ( ) correspond to the following:
    a. (Label.C) is the numeric label of the function being specified.
    b. (Label.F) are the numeric labels of the following nodes of this function COUNTER.
    c. (Quant.) is the quantity multiplier (the number of productive units produced by the system at the end of each cycle.)

Example:

9 FUNCTION COUNTER FOLLOWERS 11 7 QUANTITY 1

The general format for the consolidate function node is

(Label.C) FUNCTION CONSOL (No. to Con.) FOLL
(Label.F)

where

1. All underlined letters are key words and should be entered as shown.
2. Words enclosed in parentheses ( ) correspond to the following:
    a. (Label.C) is the numeric label of the function CONSOLIDATE being specified.
    b. (No. to Con.) is the number to be consolidated before the entity exits this node.
    c. (Label.F) is the numeric labels of the following nodes of this function consolidate.

Example:

3 FUNCTION CONSOLIDATE 5 FOLLOWERS 12

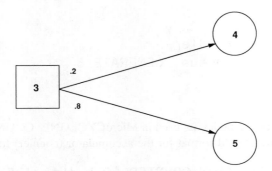

Figure C.2   Probabilistic branching.

Probablistic Arcs

COMBIs and NORMALs can be followed by a probabilistic element of two or more ARCs.

Example (See Fig. C.2).

The sum of the probabilities have to add up to 1.0.

3 NORMAL FOLLOWER 4 5 PROBABILITY .2 .8

This specifies that the entity flowing through element 3 (which is a normal) is to go to element 4 20% of the time and to element 5 80% of the time. Each time such a probabilistic branching is specified the program generates a uniformly distributed random number in the range (0,1). If the generated number is <0.2, the unit will branch to 4, otherwise it will branch to 5.

Duration Input Segment

Each task element should be accompanied with a duration set number that defines the duration type of the task and the parameters of the distribution from which the duration of the task will be sampled. MicroCYCLONE recognizes two categories of tasks based on duration-stationary tasks and nonstationary tasks.

A stationary task requires no modification of the associated duration parameters as the associated task is repeated. Unlike stationary tasks, nonstationary tasks require modification of the duration parameters that define the distribution from which duration will be sampled.

The statistical distributions recognized by the input module of the program are deterministic, uniform, triangular, beta, normal, and exponential. When using any of these distributions, the user is required to include the first three characters of the selected distributions, and then the parameters that define the statistical distribution. The following are examples of how each distribution should be defined.

Constant   DETERMINISTIC <u>PAR1</u>
    Par1   Represents the constant duration describing the duration

Uniform   UNIFORM Par1 Par2
  Par1   Represents the low value of the duration
  Par2   Represents the upper value of the duration
Triangular   TRIANGULAR Par1 Par2 Par3
  Par1   Represents the low value of the duration
  Par2   Represents the mode value of the duration
  Par3   Represents the upper value of the duration
Beta   BETA Par1 Par2 Par3 Par4
  Par1   Represents the low value of the duration
  Par2   Represents the upper value of the duration
  Par3   Represents the mean value of the duration
  Par4   Represents the variance of the duration
Normal   NORMAL PAR1 PAR2
  Par1   Represents the mean of the duration
  Par2   Represents the variance of the duration
Exponential   EXPONENTIAL PAR1
Par1   Represents the mean of the duration

Stationary Duration   The following is the general format for defining a stationary duration:

SET (set number) (distribution) SEED (seed number)

where set number is a constant number that should be associated with a defined task, distribution should be one of the distributions given above, and seed number is a constant value that should be not less than 1 or larger than 999999999.

The user may default on the value of the seed number, whereby the computer will randomly assign an initial seed number associated with the task. It should be noted that no seed number is required to be assigned when a deterministic duration is selected. The following are examples of how a distribution could be defined:

Deterministic : SET 2 DET 12
Beta   : SET 2 BET 10 15 12 .5 SEED 49483282
         or : SET 2 BET 10 15 12 .5

Nonstationary Duration   MicroCYCLONE supports two nonstationary categories. The first category includes modification of the duration parameters after the task is realized a certain number of times. After a number of realizations, the distribution parameters are changed by a specified increment value. The second category includes a change in the duration parameter after each product cycle is completed. In such situations, duration parameters will be modified using the specified increment. The following form is the general input form for both of these categories:

SET (set number) NST (distribution) Par1 Par2 SEED (seed no) where Par1 represents the value by which the distribution parameters should be incremented (this parameter should be defined in case of both categories) and Par2 represents the realization number after which the distribution parameters should be modified (this parameter should be defined in case of the first category).

The definition of the other parameters is comparable to the definitions provided in the case of the stationary durations. The following are examples of each category:

SET 1 NST DET 54 2.5 4   (deterministic)

54   Represents the duration of the task
2.5   Represents the increment by which the previous specified duration will
            be incremented
4   Represents the realization number after which the duration (i.e., 54)
            will be incremented

Category 2

    SET 1 NST BET 54 72 65.4 1.23 2.5 4 SEED 4849
    (Beta)
or SET 1 NST BET 54 72 65.4 1.23 2.5 4   ==

54 72 65.4 1.23   Represents the duration parameters corresponding to the
                   values of low, high, mean, and variance
2.5   Represents the increment by which the duration parameters
                   (i.e., low, high, and mean) will be incremented
4   Represents the realization number after which the duration
                   parameters (i.e., low, high, mean) will be incremented

Use of Database for Work Task Duration   An added feature of duration input is the compatibility of the INPUT module with the duration data file generated in BETA. The way this procedure works is as follows.

When entering the duration input and after the ''SET (set number),'' specification you don't have to enter the parameters; you can just enter REC and the name of the record. For example, enter

SET 4 REC truck

for parameters stored under name truck in the duration file through use of the BETA module and for set number 4.

To see what the file looks like, use F10 in the input menu, and it will show you the records in the STAT.DST file.

Note: In order to be able to use this facility, you must have created the records in the ''data file'' (i.e., Beta) module first. This means that when you use that module you have the opportunity to store the results in a file (STAT.DST). *Only* then can you refer to those records by name (e.g., truck in the above example) or by record number.

Equipment–Resource Input Segment

In this segment, the number of units of each resource type to be used in the network process is initialized. The types of resources include equipment (cranes, trucks), labor (concrete placing crew), or materials (a pallet of bricks). To initialize a resource, two items of information are required: (1) the number of units in the network and (2) the QUEUE node that will be the starting point for these units in the network. The header for this segment is RESOURCE INPUT, which must be typed on the first line of the segment.

The general format for the input lines is

```
(# of units) 'description' AT (Label.N.) VAR (VC)
FIX (FC)
```

where

1. All underlined letters are key words and should be entered as shown.
2. All words enclosed in parentheses ( ) correspond to the following:
   a. (NO. UNITS.) is the number of units to be initialized at this node.
   b. (Label.N.) is the numeric label of the QUEUE node where the units are to be initialized.
   c. (VC) is the variable cost associated with this unit.
   d. (FC) is the fixed cost associated with this unit.

NOTE: VAR (VC) and FIX (FC) optional.
Example:

```
4 'trucks' AT 8 VARIABLE 10.0 FIXED 25.5
```

The variable costs are the hourly costs of the specified resource based on actual operation, (fuel, oil, labor, etc.). The fixed costs are the costs, converted to an hourly basis, that are incurred regardless of whether the item is in operation (depreciation, maintenance, etc.). Variable costs apply primarily to equipment operation. Listing of variable and fixed costs is not a requirement for the program to run. These are optional inputs.

ENDDATA

The procedural word ENDDATA (END) is used to signal the end of the MicroCYCLONE input data. This will be the last line of data entered for the network. After the user has entered this line, a prompt for a new line will appear, such as

```
LINE 43 ?
```

At this point the user would press RETURN in order to leave the input mode.

Important

Before running the actual simulation the newly created input data must be compiled (Press F3). This must be done each time any changes or new information have been entered in the "Edit Input" mode. The compiler will check the entire network to ensure that no basic logic rules have been broken. Errors in network logic will be listed on the screen at the completion of the "compile" run. When any errors have been corrected, the "compile" function must be activated again.

Example Network File

An example network is shown in Figure C.3. This example shows an earth-hauling operation. This network file is taken directly from the CYCLONE model developed in Figure 6.3, Chapter 6, of this text. Probabilistic arcs and breakdown activity have been added to the basic model.

SIMULATION MODULE

The reason for modeling a production system is to examine the interaction between flow units, determine the idleness of productive resources, locate bottlenecks, and estimate the production of the system as it is designed. The SIMULATION MODULE is used in MicroCYCLONE in order to handle complex systems. During simulation, the initialized resources are moved around the network according to the durations of the work tasks. This movement is controlled by the use of a simulation clock that keeps track of the simulation time. This simulation time is based either on constant durations or on random times generated for each work task. As the simulation time advances, flow units move from one work task to another. As the resources move in a cyclical fashion through the network, certain statistics such as productivity (e.g., cubic yards of concrete per hour) and delays are gathered. In addition to the simulation clock, a counter keeps track of the number of times the flow units have cycled through the network. The simulation process stops when the simulation clock reaches the time LENGTH defined in the General Input or the counter reaches the number of CYCLES, also defined in General Input. When these two parameters are defined, it is necessary to keep in mind that there should be a sufficient number of cycles to get beyond the transient period associated with startup operations and into steady-state production.

The SIMULATION MODULE is entered through the MicroCYCLONE PROCESSING MENU which is obtained by choosing the "RUN CYCLONE" option from the main menu. The module for the monochrome display will show only the chronologic listing, whereas graphical simulation will present productivity curves.

Summary Dump of Network Logic

This listing will be displayed if the simulator encounters a problem with the model. It can also be accessed from the menu and consists of two lists describing the logical relationship of the network elements. The first list contains a description

PROCESS: demo

                        *** NETWORK FILE ***

```
LINE   1 :  NAME SPREADING LENGTH 300000 CYCLE 30
LINE   2 :  NETWORK INPUT
LINE   3 :  1 QUEUE 'GROUND AVAILABLE' GEN 30
LINE   4 :  2 COMBI SET 1 'STOCKPILE SOIL' PRE 1 3 FOLL 3 4
LINE   5 :  3 QUE 'DOZER IDLE'
LINE   6 :  4 QUE 'SOIL STOCKPILE AVAILABLE'
LINE   7 :  5 QUE 'EMPTY TRUCK AVAILABLE'
LINE   8 :  6 QUE 'FRONT END LOADER IDLE'
LINE   9 :  7 COMBI 'LOAD A TRUCK' SET 2 PRE 4 5 6 FOL 8 18
LINE  10 :  8 NORMAL 'HAUL' SET 3 FOL 9
LINE  11 :  9 QUE 'LOADED TRUCK QUEUE'
LINE  12 :  10 COMBI 'DUMP A LOAD' SET 4 PRE 9 12 FOLL 11 12 13
LINE  13 :  11 NOR 'EMPTY TRUCK RETURN' SET 5 FOL 5
LINE  14 :  12 QUE 'DUMP SPOTTER IDLE'
LINE  15 :  13 QUE 'DUMPED LOADS AVAILABLE'
LINE  16 :  14 QUE 'DUMP DOZER IDLE'
LINE  17 :  15 COM 'SPREAD DIRT' SET 6 PRE 13 14 FOL 16
LINE  18 :  16 FUN COUNTER 'ONE LOAD OF TRUCK SPREADED' QUANTITY 1 FOL 14
LINE  19 :  18 NORMAL SET 8 FOL 6 19 PROBABILITY 0.8 0.2 SEED 9999991
LINE  20 :  19 NORMAL 'BRK DWN' SET 7 FOLL 6
LINE  21 :  DURATION INPUT
LINE  22 :  SET 1 0.5
LINE  23 :  SET 2 BETA 0.94 9.33 2.09 1.2 SEED 1512
LINE  24 :  SET 3 BETA 3.67 10.17 5.65 1.62 SEED 9343
LINE  25 :  SET 4 0.5
LINE  26 :  SET 5 BETA 2.67 9.82 4.52 1.33 SEED 2356
LINE  27 :  SET 6 4.0
LINE  28 :  SET 8 0
LINE  29 :  SET 7 60.0
LINE  30 :  RESOURCE INPUT
LINE  31 :  1 'GROUND' AT 1
LINE  32 :  1 'DOZER' AT 3
LINE  33 :  1 'F.E.L.' AT 6
LINE  34 :  7 'TRUCKS' AT 5
LINE  35 :  1 'DUMP SPOTTER IDLE' AT 12
LINE  36 :  1 'DUMP DOZER' AT 14
LINE  37 :  ENDDATA
```

Figure C.3 MicroCYCLONE input file (demo).

of each work task element (COMBIs and NORMALs) and functions. The list for a typical network appears as follows:

NETWORK LOGIC DUMP REPORT

NODE 2 COMBI PRECEDERS 1 9 FOLLOWERS 1 3
NODE 3 FUNCTION FOLLOWERS 4
NODE 4 NORMAL FOLLOWERS 5
NODE 6 COMBI PRECEDERS 5 7 FOLLOWERS 7 8
NODE 8 NORMAL FOLLOWERS 9
HIT RETURN FOR DISPLAY OF QUES.

The second list describes each QUEUE node. Again from the example network:

QUE CONTENT DUMP REPORT

```
QUE 1 HAS 0 UNITS. IT IS INITIALIZED WITH 1 UNITS
QUE 5 HAS 0 UNITS. IT IS INITIALIZED WITH 0 UNITS
QUE 7 HAS 0 UNITS. IT IS INITIALIZED WITH 1 UNITS
QUE 9 HAS 0 UNITS. IT IS INITIALIZED WITH 4 UNITS
HIT RETURN TO CONTINUE WITH SIMULATION
```

These two lists can be used to double-check the network prior to continuing with simulation.

DO YOU WANT A PRINT OF SIMULATION (Y/N) ?

This output is a duplicate of the simulation chronologic list that will be displayed on the screen. If a printout is desired, the user should ensure that the print is on before continuing. Once "Y" or "N" has been entered, simulation of the process network begins.

Graphical Scale Adjustment

Before the start of the actual simulation, the user has the option of modifying the scale for the productivity curve. The default values are:

- Productivity of *50 cycles/hr*
- Total duration of *480 min*
- *No* hardcopy while simulating

If the actual simulation would not fit on the screen on the basis of the scales selected, the scale will be automatically adjusted to keep the entire graph on the screen. Adjustments will *not* be made to increase the size of a small graph. This must be done by the user, through adjustment of one or both scales. Figure C.4 represents a graphical output of a simulation run.

REPORT MODULES

The report modules are entered from the MicroCYCLONE PROCESSING MENU by choosing option "F3." Two report modules are available, one for monochrome tabular screens and one for graphics only. The following discussion pertains to the tabular reports only. The idleness and production figures in this text were generated using the graphical report capability of MicroCYCLONE.

There are five different types of reports available. Each of these reports describes a different category of statistics or results that have been produced by the most recent simulation of the process.

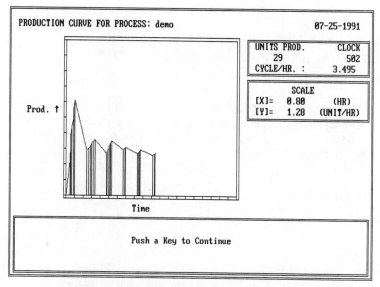

PRODUCTION CURVE FOR PROCESS: demo   07-25-1991

UNITS PROD.	CLOCK
29	502
CYCLE/HR. :	3.495

	SCALE	
[X]=	0.80	(HR)
[Y]=	1.28	(UNIT/HR)

Prod. ↑

Time

Push a Key to Continue

Figure C.4 MicroCYCLONE production curve (demo).

Report Module Menu

1. REPORT BY ELEMENT
2. CYCLE MONITORING
3. PRODUCTION BY CYCLE
4. PROCESS REPORT
5. NETWORK LOGIC REPORT
6. SYSTEM MENU

One feature of the report format is that reports that cannot fit on the screen all at once are displayed as pages. At the bottom of the screen the program will display:

M-MAIN MENU   R-REPEAT LAST PAGE   C-CONTINUE

This allows the user to continue for further information (enter ''C''), go back and review previous information (enter ''R''), or return to the report module menu (enter ''M'') without looking at the remaining data. Each report in the report module will be discussed separately.

Report by Element

The first option is the report by element. This report is selected by entering the number ''1.'' After the RETURN is pressed, the program will display the following submenu:

Report by Element

1. SORTED BY LABEL IN INCREASING ORDER
2. USER SPECIFIES NODES TO BE INCLUDED
3. PRECEDERS AND FOLLOWERS

The information in this report can be displayed in three different formats. The first is a complete listing of information for all network nodes. First, the work task node reports (COMBIs and NORMALs) are displayed, then the QUEUES, then the FUNCTION nodes.

Report on Work Tasks

COUNT is the number of times that the work has been activated.

MEAN DUR is the mean length of time that it took for that work task to be accomplished. For constant-duration work tasks, this will be the same time as that defined in the parameter set for that node. For nonconstant durations it will be the mean of the times generated by the program.

AR. TIME is the average time between the arrival of flow units at that node.

AVE. NUM presents the average number of units positioned (actively involved) at that particular work task at any time.

%BUSY is the percentage of time the work task was in operation.

Report on Queues

AVE. WAIT shows the average amount of time that a unit was waiting (or "idle") in the queue before moving to the following work task. It is a key indicator of the delays that occur in the process.

AVG. UNIT presents average number of units at any one time that occupied the QUE node.

UNITS END is the number of units in the QUE waiting to move on to the next work task at the end of the simulation.

Report on Functions

In the FUNCTION report, *COUNT* indicates the total number of flow units that have gone through the counter node. This, by definition, is equal to the number of "cycles" the network has passed through. *STATISTICS BETWEEN* is a measure of the average time between flow units passing through the counter. The *FIRST* statistic is the time at which the first flow unit passed through the counter.

The second option in the "Report by Element" report enables the user to obtain the same reports as the first option (described above). The only difference however is that the reports are generated for those nodes the user specifies and not for all nodes as in the first option.

The user must enter the "number of nodes to be included" with the report at the prompt. This corresponds to the number of nodes for which the user wishes to generate reports. Then one enters the labels of the nodes and the reports will be generated.

The third option in the "Report by Element" report is to make a detailed analysis of a particular COMBI. After the user has entered the number "3," the program asks for the number of the COMBI to be examined.

```
DETAILED ANALYSIS OF COMBI ACTIVITY
ENTER THE LABEL OF THE DESIRED ACTIVITY:
```

The user enters the COMBI number and the program displays the type of report discussed above. This report provides the statistical output for that node as well as for all preceding and following nodes. This option can be used only with COMBI nodes. If the user enters the number of a node that is not a COMBI, the program will return an error message:

```
NODE < node number > IS NOT A COMBI
```

Cycle Monitoring Report

This report allows the user to monitor at which points in time each work task has been realized.

T-NOW shows the time at which that particular cycle of the work task began.

COUNTER indicates how many times the node has been occupied. The complete output for this report would show all cycles for each work task until the COUNTER first reached the number of cycles specified in the process input phase.

Production by Cycle Report

This report shows the production in units per hour per cycle. The report shows the cycle number, the simulation time when that cycle was completed, and the cumulative productivity at that time. The effect of the startup (or transient) period on the system is displayed and the gradual progress into steady-state production can be observed. It may suggest the need to simulate more cycles in order to reach the steady-state phase.

Process Report

This report is a summary of production results in quantities per hour. The user will be asked to enter an operating factor by defining productive minutes per hour. The productive minutes per hour will vary with particular job conditions. Factors affecting productivity include weather, type and availability of equipment, management effectiveness, and work-face access. Without detailed knowledge of job conditions, it is standard practice to assume that, on the average, only 50 min of each hour is fully productive time.

If the model being simulated is based on an actual construction project, for which activity times have been taken in the field over an extended period, then the full 60 minutes can be used.

RUN LENGTH is the total simulation time.

NUMBER OF CYCLES shows the actual number of simulated cycles.

The rest of the information displayed is self-explanatory. Total cost and the cost per unit will be zero when no cost information was included in the network input. If variable and fixed costs have been included when entering the RESOURCE INPUT, the cost results will be displayed here.

Network Logic Report

This report consists of two sections: (1) a list of all work tasks and FUNCTIONS in numeric order, showing the preceding and following elements for each one, and (2) a list of all QUEUES, with the number of flow units initialized at each QUE at the beginning of the simulation and the number of units remaining in the QUE at the end of simulation.

Statistical Analysis of Input Data (Beta Module)

The beta distribution program is a supportive module for the MicroCYCLONE simulation package. The program is an interactive, menu-driven software module that accepts data from the user—via either keyboard or file—in either of two different ways: (1) grouped data in the form of frequencies for a particular interval of durations or (2) ungrouped (raw) data in the form of observations recorded at points of time. In the latter case the data are grouped using Sturges' rule and transformed into type (1) grouped data. The original data are, however, kept intact.

The sample data histogram is shown on the screen, and a beta distribution is selected to best fit the data by the method of moment matching. The beta distribution selected is plotted on the same screen superimposed on the histogram of the original data in one of two forms: (1) beta histogram corresponding to the same intervals or (2) both beta histograms and beta curve.

The user is able to check the fit of the selected distribution visually, or by requesting a chi-square (χ^2) goodness-of-fit test. The beta parameters as well as the results of the chi-square test are all displayed in the same screen with the histograms and beta curve. The user, if satisfied with the selected distribution, can save the parameters directly into a database that can be accessed by Micro-CYCLONE on request. This facilitates keeping track of all analyzed cases. The parameters can be accessed by referencing a record name (or number) specified during the analysis procedure. Help facilities in the input module allow recalling the database file for quick revision while entering the MicroCYCLONE input file.

Problems

Chapter 1

1.1. Study the operations of a local construction company and identify the levels of hierarchy described in this chapter. Identify:

    a. Two projects in progress.

    b. For each project, three activities in the project schedule.

    c. For each project, two operations that are significant from a project point of view.

    d. Three construction processes on each project.

    e. Two work tasks in each of the processes identified.

1.2. Identify three construction processes or technologies that rely on repetition to achieve improved production.

1.3. Describe a technique (such as the cable-supported bridging technique) that you feel can lead to repetition and thereby improve construction productivity.

1.4. Describe industrial processes you are familiar with in which a physical model is used to improve planning and design of the end product. Similarly, describe an industrial process in which an abstract model (mathematical, analog, etc.) is used to analyze and evaluate an end product. (*Hint:* The beam equation for analysis of a structural element can be thought of as a mathematical model of the beam.)

1.5. What is the difference between a deterministic and a probabilistic model?

1.6. What is the significance of the balance point in a two-cycle deterministic system?

1.7. Describe a production or materials handling system in which three cycles interact forming a three-link-node system. Can you think of a system which can be modeled as a four-link-node system?

1.8. Identify the various work tasks and processes involved in the following material handling situations.

    a. Transit mix concrete delivery to a high-rise city building material hoist skip with floor hoppers and buggy distribution at floor level.

b. The delivery of reinforcing steel to a ground site area with subsequent manufacture of column cages and insertion into column forms on the working floor of a high-rise building.

c. Spoil removal from a tunneling project.

d. A quarry and crushing plant operation.

1.9. An earthmoving operation uses scrapers for haulage and a pusher dozer for loading operations. Develop a detailed list of work tasks for the operation.

1.10. Several tractor scrapers are being push loaded by a ''pusher'' dozer. The average ''push'' time is 1.00 min. The average return time (to get into position to push the next scraper) is 0.4 min. After the loading, the average travel to dump time is 6 min and the average return time is 4.5 min for the scrapers. Dumping material requires 0.25 min. Each scraper holds 30 loose cubic yards.

a. What is the balance point of the system?

b. If 6 scrapers are supported by 2 pushers, what would you predict as the system production?

c. Is the system operating above or below the balance point?

CHAPTER 2

2.1. Consider the road job described in Figure 2.1, which consists of 14 road sections to be completed. The objective chart for this job is given in Figure P2.1.1. Assume that each month consists of 20 working days on the average. The program chart for this process is given in Figure P2.1.2. Calculate the line of balance for the study representing the beginning of month 8.

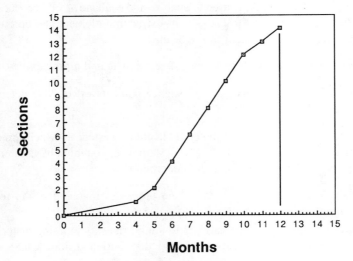

Figure P2.1.1   Objective chart.

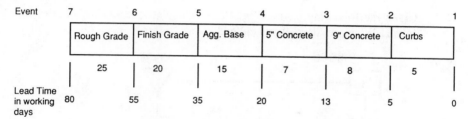

Event	7	6	5	4	3	2	1
		Rough Grade	Finish Grade	Agg. Base	5" Concrete	9" Concrete	Curbs
		25	20	15	7	8	5
Lead Time in working days	80		55	35	20	13	5	0

Figure P2.1.2   Program chart.

2.2.   Figure P2.2 shows the line-of-balance objective chart for a 10-story building. Each floor is divided into four sections (*A, B, C, D*). The production for a typical section is shown below the objective. Calculate the LOB values for control points 1–8 for week 5 (200 hr). Give LOB values in numbers of floor sections.

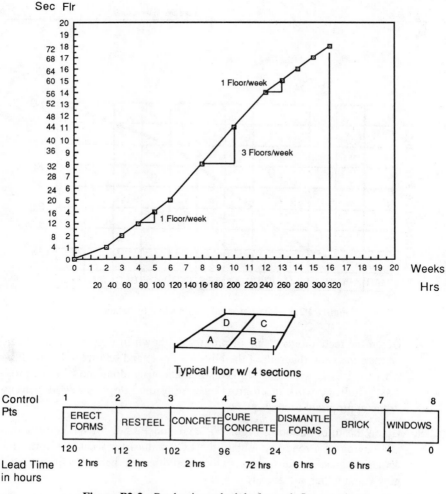

Typical floor w/ 4 sections

Control Pts	1	2	3	4	5	6	7	8
	ERECT FORMS	RESTEEL	CONCRETE	CURE CONCRETE	DISMANTLE FORMS	BRICK	WINDOWS	
	120	112	102	96	24	10	4	0
Lead Time in hours	2 hrs	2 hrs	2 hrs	72 hrs	6 hrs	6 hrs		

Figure P2.2   Production schedule for each floor section

2.3. Given the following charts (Fig. P2.3), calculate the LOB quantities for July 1. What are these charts called?

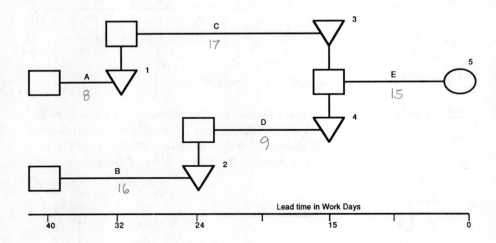

Lead time in Work Days

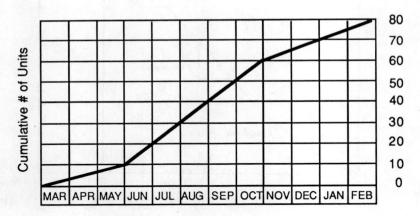

Figure P2.3 Charts for calculating line of balance.

2.4. Given the technologic "line of progress" shown in Figure 2.19, indicate on the schematic diagram of the building floors and sectors in Figure P2.4 what work tasks are being worked simultaneously along the "line of progress." Indicate work in progress by writing the title of the work activity in the appropriate sector square.

2.5. Given the information in Figures P2.5.1 and P2.5.2 for the construction and paving of a roadway, obtain the LOB values for week 10. From the actual progress profiles given, which activities are behind schedule (lagging), and which are ahead?

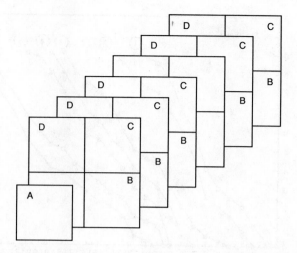

Figure P2.4 Schematic of building floors and sectors.

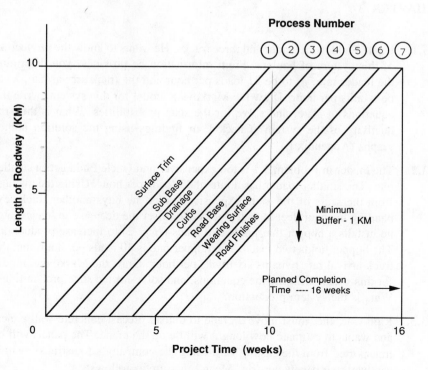

Figure P2.5.1 Planned program for highway construction.

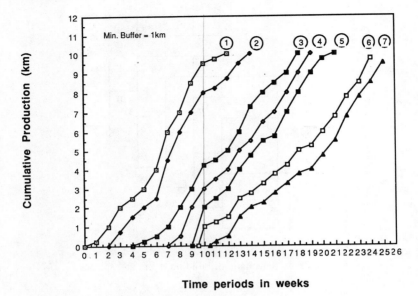

Time periods in weeks

Figure P2.5.2 Actual progress to completion.

CHAPTER 3

3.1. Uncle Fudd has a loader and three trucks. He wants to know the production of this system of haulers. From information he provides, you determine the loader rate, μ, to be 21 loads per hour and the truck arrival rate, λ, to be 6 units per hour. Draw the Markovian model for this system, write the equations of state, and solve for the state probabilities. What is the productivity of the system? Check your findings using the solution nomographs (Appendix B).

3.2. The loader in Problem 3.1 is too expensive, and Uncle Fudd gets a smaller one. The smaller loader has a rate of 12 loads per hour. Using the average from the scale of the more expensive loader, he buys another truck, expanding his haul fleet to four trucks. To offset the decrease in loader rate, he installs a hopper that holds two loads ($H = 2$) to increase production. The hopper holds two loads. The hopper rate is 30 loads per hour, and the truck arrival rate remains six trucks per hour. Draw the Markovian model for this situation, write the equations, and solve for the state probabilities. What is the system production?

3.3. Ripp Co., Inc. must move a crane to unload precast concrete wall panels and wants to estimate how long it will tie up the crane. The panels will be transported from the casting yard using the company's four trucks, which can haul two panels per trip. Mean times are as follows:

a. 24 min to unload two panels

b. 19 min to travel to casting yard (empty)

c. 10 min to load two panels.

d. 31 min to travel (unloaded) to the job site.

Using a queueing formulation, determine how many 8-hr shifts with the crane will be required to unload 100 panels.

3.4. Five masons are supported by two laborers. The laborers carry bricks to the masons at a rate of one packet containing 10 bricks every 2.2 min on the average. The placement rate by the masons averages 10 bricks every 6.2 min. There is no space on the scaffold for temporary storage.

    a. Solve the system of state equations and determine the productivity of the system.

    b. Using the situation as described above, assume that one laborer is now on vacation and the scaffold has been modified to allow stacking of brick packets on it. Calculate the hourly production of this modified system and hence determine the stack storage required (and therefore the laborer overtime) to ensure that the single laborer can support the masons during their 8-hr work shift.

3.5. The CONJOB Construction Co. is completing a job in which pouring of the basement slab was delayed until the building was nearly complete. The pour will be made by wheelbarrows filled from a bin that is fed from a chute (elephant trunk) through one of the windows. The bin holds a maximum of 6 cubic units of concrete. The mixer on site is capable of producing 72 cubic units of concrete per hour, and the wheelbarrow will be filled with 2 cubic units per trip. The rate of loading the wheelbarrows from the bin is one every 20 sec. The floor pour requires 360 cubic units of concrete. There are five wheelbarrows available, and the travel cycle is 8 min. Construct the Markovian model for this system and write the equations of state. Calculate the productivity of the system using the appropriate nomograph (Appendix B).

3.6. A lake-front cofferdam is to be constructed to expedite the construction of a pier. The coffer will be constructed using a double row of sheet pile filled with granular material to form a coffer wall. The lateral stability for the sheet pile walls will be developed by tying them back to one another. The tie installation is to be accomplished using two cranes and eight installation crews. The mean time for initial installation of the rods is 10 min. During this time the rods are spot-welded in place and the crane is required to hold them in position. Following this, final welding of the rods is completed, requiring the crew for an additional 25 min, on the average. The crane, however, is free to work on other rods during this period. How many rods can be installed per hour?

3.7. The field data given below is similar to that shown in Table 3.1. Using a form similar to the one shown in Table 3.2, calculate mean and values for each truck. For the system described, determine the system productivity in truck loads per hour. The values given are in the format $M \cdot S$ where M represents minutes and S seconds. A single loader is utilized. What is the mean length of the truck waiting line?

Truck number	1	2	3	4	5	6	7	8
Arrive	0.00	0.00	0.00	6.10	7.05	8.45	14.01	14.52
Load	0.18	5.28	3.16	9.39	7.55	11.31	14.36	16.31
End of load	2.03	7.28	4.50	11.01	9.21	13.17	16.05	17.55

Truck number	1	3	9	5	2	4	6	7
Arrive	15.40	20.29	22.12	24.38	23.20	27.56	33.10	34.27
Load	18.24	21.38	24.18	29.20	26.51	31.50	33.50	36.33
End of load	19.43	23.50	26.33	30.46	28.50	33.49	36.15	38.02

Truck number	8	1	3	4	2	5	4	6
Arrive	35.00	36.04	41.40	41.30	45.46	46.35	51.41	55.10
Load	38.28	40.19	41.59	45.15	48.33	50.35	55.24	57.25
End of load	40.01	41.40	44.00	47.11	49.44	54.55	56.53	58.50

Truck number	7	8	1	3	2	5	4	6
Arrive	57.10	57.30	57.40	62.51	63.24	69.13	73.00	73.40
Load	59.18	61.22	63.35	65.53	68.39	71.25	73.38	75.43
End of load	60.53	63.08	65.20	68.11	70.49	73.10	75.18	77.50

Truck number	8	3	2	7	5	4	6	
Arrive	76.36	85.51	86.30	87.01	88.08	92.42	93.44	
Load	97.32	100.21	104.45	106.27	108.09	0.00	109.58	
End of load	99.51	102.22	106.02	107.39	109.28	0.00	111.42	

Truck number	8	3	7	5	6	8	2	
Arrive	112.12	121.00	121.20	124.33	126.15	130.05	134.06	
Load	115.27	123.23	126.35	128.42	131.54	133.44	135.46	
End of load	116.11	125.58	128.04	130.23	133.01	135.12	137.11	

Truck number	3	5	8
Arrive	142.10	147.08	148.03
Load	142.38	147.49	149.50
End of load	144.41	149.25	151.05

3.8. The following tables represent field data on haul and load times. Calculate the production of this system deterministically and using the appropriate queueing nomograph in Appendix B. Develop a plot of the system production both deterministically and using the queueing approach for systems from 1 to 10 trucks. Why are the plots different?

Loader Time in Minutes		Back Cycle in Minutes	
Interval	Frequency	Interval	Frequency
0–0.5	2	4.0–4.5	4
0.51–1.0	5	4.51–5.0	10
1.01–1.5	10	5.01–5.5	28
1.51–2.0	20	5.51–6.0	32
2.01–2.5	25	6.51–7.0	20
2.51–3.0	25	7.01–7.5	4
3.01–3.5	10	7.51–8.0	2
3.51–4.0	2		100
4.01–4.5	1		
	100		

3.9. **a.** A loader loads a fleet of six trucks engaged in earthmoving. It takes the loader 2.5 min to load one truck. The back-cycle time for one truck is 6.0 min. Construct the Markov model for this system and write the state equations. Do not solve these equations, but determine the production of the system using nomographs.

b. Assume the fleet of trucks has been reduced to four and a hopper has been introduced into the system. The hopper holds two truck loads, and it takes 0.5 min to dump to a truck. Construct the Markov model for this system. Calculate the production of this system using nomographs.

3.10. Five masons are supported by one laborer. The laborer carries bricks from a stockpile and places 10 brick packets on the scaffold. The placement and carrier times observed in the field are given in the following tabular list:

10 Brick Placement Times		Carrier Cycle Time (10 bricks)	
Interval	Frequency	Interval	Frequency
3.0– 4.0	12	0 – .5	2
4.0– 5.0	15	.5–1.	8
5.0– 6.0	18	1. –1.5	16
6.0– 7.0	21	1.5–2.	20
7.0– 8.0	16	2. –2.5	22
8.0– 9.0	12	2.5–3.	16
9.0–10.0	6	3. –3.5	9
		3.5–4.	4
		4. –4.5	3

The design of the system allows for stacking of 10 brick packets on the scaffold. The number of stack positions is four. Calculate the productivity of the system (use nomographs in Appendix B). Assume pickup time from the stacks to be 1.0 min.

a. What is the production of the five-mason system?

b. What is the balance point of the system?

c. Is the system operating above or below the balance point? Discuss the meaning of this.

d. If the number of masons can be varied up to eight, what is the optimal number of masons supported by the single laborer? Assume that the masons make $16.00 per hour and the laborer earns $8.60 per hour.

(*Note:* The system is as shown schematically in Figure 10.6.)

CHAPTER 4

4.1. Define the following terms as used in MPDM:

a. Production unit

b. Production cycle

c. Leading resource

4.2. MPDM was conducted to improve the concrete placement process for a building deck slab. The process is based on pumping the concrete. The collected data are shown in the table below.

PRODUCTION CYCLE DELAY SAMPLING								
Page 1 of 1 Date:					Unit:			
Method:					Production unit: None			
Prod. Cycle	Prod. Cycle time (sec)	Envi. Delay (sec)	Equip. Delay (sec)	Labor Delay (sec)	Material Delay (sec)	Manag. Delay (sec)	Notes non–delay	Minus Mean Non–Dealy Time
(1)	(2)	(3)	(4)	(5)	(6)	(7)	(8)	(9)
1	900						Non–delay	100
2	1500		40%		60%			700
3	750						Non–delay	50
4	750						Non–delay	50
5	1000					*		200
6	1200		*					400
7	950			*				150
8	950			20%		80%		150

MPDM   PROCESSING

Page 1 of 1   Date: 08/14/89   Unit: Seconds
Method: Concrete placement   Prod. Unit: Truck loads

Units	Total Production Time	Number of Cycles	Mean Cycle Time	Sum[1(cycle Time)-(N.D. Cycle Time)1]/n
A.Non-Delayed Production Cycles	2400	3	800	
B.Overall Production Cycles	8000	8	1000	

DELAY   INFORMATION

	Delays				
	Manag.	Equip.	Labor	Mater.	Evir.
C.Occurrence	2	2	2	1	0
D.Total Added time	320		180	420	0
E.Probability of occurrence	0.25	0.25	0.25		0
F.Relative severity		0.34	0.09	0.42	0
G.Expect % delay time per prod. cycle	4	8.5		5.25	0

Complete the MPDM processing sheet given above.

4.3. You are given the MPDM processing sheets shown below.
Determine the following (show your calculations):
   a. Ideal productivity
   b. Overall method productivity
   c. Ideal cycle variability
   d. Overall cycle variability

```
MPDM   PROCESSING

Page 1 of 1      Date: 08/02/90        Unit: Minutes
Method: Concrete Placement            Prod. Unit: Truck loads
```

Units	Total Production Time	Number of Cycles	Mean Cycle Time	Sum[l(cycle Time)-(N.D. Cycle Time)l]/n
A. Non-Delayed Production Cycles	43	2	21.5	1.5
B. Overall Production Cycles	224	5	44.8	23.9

```
                    DELAY   INFORMATION
```

	Delays				
	Manag.	Equip.	Labor	Mater.	Evir.
C. Occurrence	1	0	3	0	0
D. Total Added time	97	0	19.5	0	0
E. Probability of occurrence	0.2	0	0.6	0	0
F. Relative severity	2.165	0	0.145	0	0
G. Expect % delay time per prod. cycle	43.3	0	8.7	0	0

4.4. Observed cycle times for a crane and bucket concrete operation are in Figure P4.4. What would you estimate as the productivity of this system using MPDM?

4.5. Observed cycle times for a concrete finishing operation are given in Figure P4.5. What would you estimate as the productivity of this system using MPDM? How much does each delay source contribute to the loss in production? If you were in charge of this operation how would you improve it?

Production Cycle	Cycle Time (seconds)	Environment Delay	Equipment Delay	Labor Delay
1	73			
2	78			
3	65			
4	126		✔	
5	76			
6	150			✔
7	90			
8	64			
9	300	✔		
10	85			
11	72			
12	122			✔
13	116			✔
14	202		✔	
15	75			

Figure P4.4 MPDM data sheet.

Production Cycle	Cycle Time (seconds)	Environment Delay	Equipment Delay	Labor Delay
1	103	✔		
2	72			
3	65			
4	140			
5	82			
6	134		✔	
7	85			
8	220			✔
9	128			✔
10	66			
11	440		✔	
12	75			
13	305	✔		
14	155		✔	
15	72			

Figure P4.5 MPDM data sheet.

4.6. The observed cycle times for a brick placement operation are given below. Using MPDM:

  a. Calculate the ideal, overall, and method productivities of this operation.

  b. Compute the relative severity and expected percent of delay for each of the three types of delays given in this process.

Production Cycle	Cycle Time Seconds	Management Delay	Equipment Delay	Labor Delay
1	130			
2	166			
3	114			
4	126			
5	120			
6	134			
7	258		x	
8	192			
9	218			x
10	120			
11	133			
12	122			
13	112			
14	116			
15	142			
16	140			
17	124			
18	194			
19	186			
20	258		50%	50%
21	232		50%	50%
22	230		50%	50%
23	340	50%	50%	
24	176			
25	252			x
26	282			x
27	306	x		
28	168			
29	290			x
30	182			
31	192			
32	200			
33	230		x	
34	206			

4.7. Select a heavy-construction job site in the area and visit the site.

a. Draw a physical plan of the site (i.e., a site layout diagram).

b. What materials handling and processing systems can you identify?

c. Select an excavation, grading, or similar earth construction process for further study.

d. Conduct a MPDM study of one of the processes to determine
   (1) Ideal production
   (2) Actual production
   (3) Sources of delay
   (4) Magnitude of delay from each delay source
   (5) Actions you would take to improve process

e. Discuss with the job superintendent the rationale for setting up the job and the basis for the fleet of equipment, size of crews, etc. being used.

     f. Compare the productivity established with a standard reference and comment.

4.8. Discuss how you might use the MPDM (method productivity delay model) in a real-world situation. What is the significance of the method variability? How does it differ from standard deviation, and why isn't standard deviation used?

CHAPTER 5

5.1. Eight (8) masons are supported by two (2) laborers. A mason removes one 15-brick packet from a stack position on the scaffold in about 1 min and places it in about 11 min. The laborers start supplying a brick packet to a stack position, when the preceding packet has been removed by the mason. Four stack positions are available near the working area of the masons. The average time for supplying one brick packet is 2 min, and 7.2 bricks are required for one square foot of the wall.

     a. How much is the production per 60-min hour for each of the three links of this three-link node system?

     b. How long will it take to place 500 ft^2 of running wall?

     c. How many space units and laborers would balance the eight masons?

     d. What are the idle times (in percent) of the masons and the laborers in the steady state, if deterministic behavior is assumed?

5.2. Three trucks haul material from a loading tower to an airfield construction job. At the fill (i.e., airfield) a spotter shows each truck where to dump. After the trucks dump material, they return to the loading tower. A front loader is used to load the tower. The tower can hold up to three loads. Draw a circle-and-square (CYCLONE) diagram of this system. Indicate all cycles and where the units in the system would be initialized at time zero.

5.3. A number of tractor scrapers are being push-loaded by one "pusher" dozer. The average "push" time is 1.5 min. After pushing a scraper the dozer returns to the "push point" to await another scraper. If another scraper is available to push, the dozer engages the booster bar and resumes pushing. The return time to push point is 1.0 min. After loading the average travel time for a scraper to the fill area is 15.0 min. The average empty return time is 10.0 min. Dumping the material at the construction site takes 1.0 min. Draw a model of this system using circle-and-square notation (include a COUNTER). What is the balance point of the system?

CHAPTER 6

6.1. Brick pallets are picked up from the supplier and transported by truck to the job site, where they are off-loaded and stockpiled. Draw a model of this process similar to the one developed for the earth-hauling operation shown in Figure 6.3.

6.2. Visit a job site and select a process for investigation. Draw a schematic diagram of the process and/or site layout. Identify the major flow units in the process selected and list the active and waiting states through which each unit passes.

6.3. Develop CYCLONE models similar to that of Figure 6.3 for the following labor and craft situations:
   a. The erection of column formwork by a carpenter and laborer crew
   b. The field operation of a drilling machine
   c. The placement of concrete in a slab using buggies and vibrators with a concrete crew made up of laborers, cement finishers, and a supporting ironworker and carpenter

6.4. Steel sheet piles 12 ft long are being driven using a double-acting compressed-air hammer. The steel is positioned initially using a driving template. A mandrel is placed on the pile once it is in position and driving commences. When the pile has been driven 8 ft, the hammer and mandrel are widthdrawn and another 12-ft section is welded to the first. After this, driving continues. This process continues until four sections have been welded to the original and driven. The last section is trimmed to a uniform elevation using a cutting torch. The next section of wall is started by positioning the initial sheet pile in the template with interlock to the just-completed drive. Assume that the sheets are stacked initially in a stockpile location. Identify the active and waiting states through which each sheet pile must pass from beginning to end of the process. What resource units might constrain the movement of the pile from stockpile to final driven location?

6.5. A number of scraper pans are repaired by a maintenance crew when they break down. After repair, the scrapers continue to run until they break down again. The times to repair a scraper pan (as developed from field data) and the times that a typical pan runs until it breaks down again are shown on the next page. Draw a model of this breakdown repair system using CYCLONE notation. What is the balance point of the system?

Breakdown Data		Run Time Data	
Interval (hr)	Frequency	Interval (hr)	Frequency
0 –0.99	13	0–2	4
1.0–1.99	10	2–4	19
2.0–2.99	8	4–6	47
3.0–3.99	8	6–8	79
4.0–4.99	6	8–10	76
5.0–5.99	5	10–12	67
6.0–6.99	3	12–14	54
7.0–8.0	3		

6.6. Transit mix trucks are hauling to a high-rise building construction where a floor slab pour is in progress. The concrete is lifted from street level using a tower crane–concrete bucket system. Two concrete buckets are used for the lift. After being lifted, the concrete is stored temporarily in a storage hopper at the floor level of the pour. The concrete is removed from the hopper and carried to the placement location by rubber-tired buggies. Determine the major cycles in this system and diagram the active and idle states in each cycle showing the direction of flow of processed units. Integrate the cycles to show transfer points between one cycle and any interface it has with other cycles. What is the effect of having two concrete buckets? How big should the storage hopper be?

CHAPTER 7

7.1. What would happen to the flow unit population of the systems illustrated in Figures P7.1a–c after 10 cycles? After 30 cycles?

7.2. Simplify or correct networks of Figures P7.2a–f.

7.3. A pallet truck picks up 35 brick pallets at the vendor's location and transports them to the job site, where they are off-loaded and stockpiled as 35 individual units. Draw a simple model for this situation utilizing the CONSOLIDATE and GENERATE functions to minimize the number of units flowing in the system. Locate a COUNTER so as to measure the production rate in pallets stockpiled at the job site.

7.4. There are four columns to be poured, all of which are 24 in. in diameter and 10–12 ft in height. Therefore, each column requires approximately 1–1.5 yd^3 of concrete. Prior to placing concrete, the columns are formed

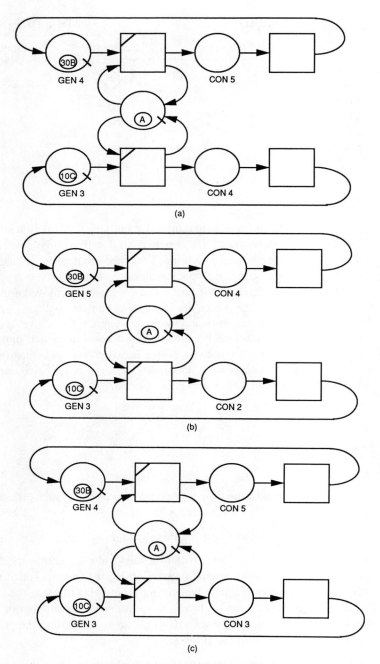

Figure P7.1 GEN–CON models.

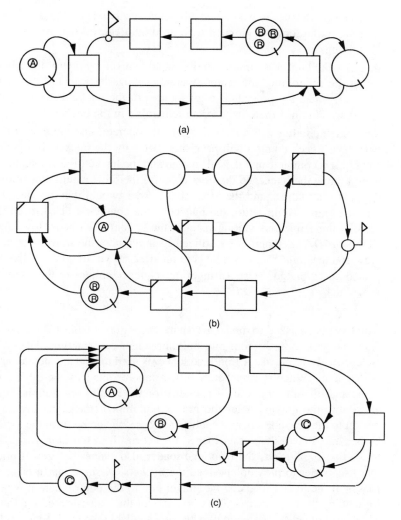

Figure P7.2 Networks for simplification.

with circular metal forms and scaffolding is assembled in position around each column form. The concrete placement sequence is as follows:

a. The crane lowers the concrete bucket (1-yd^3 capacity) into position at the transit mix truck, where a laborer positions the chute so that the bucket may be filled. The truck operator (teamster) regulates the feeding of concrete into the bucket.

b. After the bucket is filled, the crane swings the bucket to a position over one of the columns, where it is positioned into place by two laborers working on the scaffolding.

c. Once the bucket is positioned, the concrete is released and, after settling in the form, is vibrated by a third laborer working from the scaffolding.

d. After the bucket is emptied and a signal is given by one of the laborers on the scaffolding, the crane operator swings back to the transit mix truck where the procedure begins anew, or the bucket is swung to another column if there is concrete remaining in the bucket.

Since each column will hold 1–1.5 yd^3 of concrete, and the bucket holds only 1 yd^3, each column will be visited twice by the bucket. If a pour on a column is complete and 0.5 yd^3 remains in the bucket, the crane operator would move the bucket to the next column, the crew would reposition on the next scaffolding, and the remaining 0.5 yd^3 would then be placed before the crane would return the bucket to the concrete truck for refilling. Identify the active and waiting states in the system described. How can the CONSOLIDATE function be utilized to implement the movement of the crew and the crane to the second column after 1.5 yd^3 of concrete has been placed and vibrated? (Use a triggering sequence similar to that shown in Fig. 7.8.)

7.5. You have been asked to model a system where gravel ballast is transported from the ground to the roof of a building under construction. The system is described as follows. A front-end loader is used to load a concrete bucket with gravel. When the bucket is full, a laborer attaches the bucket to a crane and lifts it to the roof. A spotter is on the roof to position the bucket and dump the gravel. After the gravel is dumped, the crane lowers the bucket to the ground, where the laborer releases the crane from the lowered bucket and attaches the other loaded bucket to lift to the roof. In addition to loading the bucket, the front-end loader also shapes the gravel pile so that it is easier to pick up and load the buckets. Assume that it takes two passes of the loader to shape the gravel pile to the point where it is ready for loading the two buckets. Only one pass of the loader is required to load a bucket, however. There are two buckets available, so the loader can load one bucket while the other is on its way to the roof and back. At the beginning of the process, the gravel stock is shaped and ready to load both buckets. Draw the CYCLONE network to represent the system described above.

7.6. Draw a CYCLONE model of the following construction operation. Twenty-four columns are to be formed prior to placing concrete. An 8-ft cage of reinforcing steel is prefabricated by the steel crew prior to being lifted in place and attached to existing dowels from the column below. The steel crew attaches the cages. Carpenters also prefabricate wooden column forms. When the reinforcing is in place, the prefabricated forms are lifted

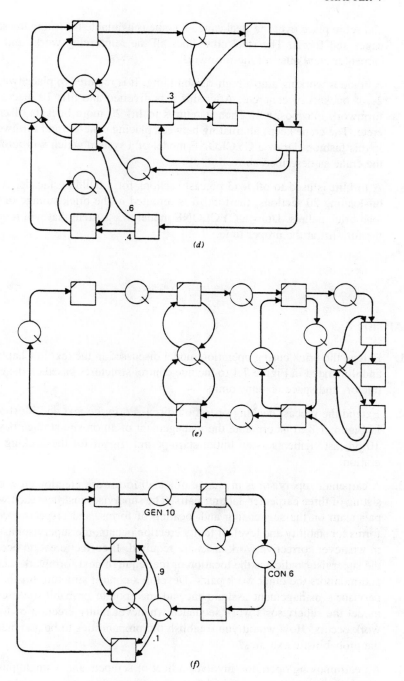

(d)

(e)

(f)

GEN 10

CON 6

and set in place over the steel cage. A crane is available for lifting the steel cages and forms. The steel crew does all the reinforcing work, and the carpenter crew does all the formwork.

7.7. A crane is working atop a high-rise building. It is required to place twenty 2-yd³ buckets of concrete. After this it is diverted and lifts 10 pieces of formwork into position. Then it returns to lift 20 more buckets of concrete. The crane keeps alternating between placing concrete and formwork in this fashion. Draw a CYCLONE model of a system which will reroute the crane as described above.

7.8. A forklift is used to off-load precast sections for a building facade. After off-loading 20 sections, the forklift is rerouted to the brick storage to off-load brick pallets. Draw a CYCLONE model of a system that will reroute the fork lift at the proper time.

CHAPTER 8

8.1. Extend the brick curing operation model discussed in the text in Chapter 7 and illustrated in Figure 7.1 to include control structures based on the cart, tractor, and space resource units.

8.2. Extend the precast decking operation discussed in the text and illustrated in Figure 7.8 to incorporate the management of an on-site storage bay of 10 precast elements as an initial start–permit trigger for the decking operation.

8.3. A carpenter supervisor is in charge of a slab formwork erection crew consisting of three carpenter–laborer pairs. The supervisor instructs each work pair team on the sequencing and locating of forms, and inspects erected forms for stability and level. If faulty erection occurs, the supervisor assists in whatever corrective work tasks are required. If correct erection occurs, the supervisor indicates the location of the required next formwork section and monitors the other work pairs. Develop a control structure for the supervisor's management assignment and incorporate probabilistic arcs to model the supervisor's work sequence whenever faulty erection of formwork occurs. How would you establish the probabilities to be assigned to the probabilistic exit arcs?

8.4. An earthmoving operation involves a fleet of scrapers and a single pusher dozer for the loading operation. Scraper operators have been instructed to bypass the pusher dozer and self-load to minimize bunching effects whenever the dozer in engaged with a scraper in the load operation and a second scraper is already available in a queue position awaiting the pusher dozer. Develop a suitable control structure to model the earthmoving operation.

8.5. The process under consideration involves the movement of precast double-T parking decks from a storage yard to the job site where a parking garage is being constructed. The double-T precast elements are moved using low-bed trucks. Two elements can be loaded on each truck. Each section is lifted into position at the job site using a tower crane. Because of the congested nature of the job site, trucks must wait in the street while being off-loaded. Only two trucks can wait at a time. When a truck has been unloaded, a signal is sent to the storage yard to release the next truck. Model the precast element described.

8.6. In Problem 8.5, assume that trucks are not held at the storage yard but move to the job site directly. At the job site, there are two off-loading positions. Incoming trucks are delayed at a holding area until one of the two off-loading positions becomes available. As an off-loading position becomes available, the next truck moves forward, off-loads, and departs. Model this slightly modified situation.

8.7. Trucks off-load at a warehouse that has four unloading bays. The nature of the location is such that trucks waiting to off-load must wait on the main street, since the access road would be blocked otherwise. A site plan is shown in Figure P8.7.1. Develop a model of this process. The model should be constructed so as to indicate periods when the access road is blocked to other traffic because of the maneuvering of trucks that are docking. For the purposes of running a computer simulation of this process, the data below is provided.

a. Truck entry time (from waiting line): 5 min

b. Truck departure time (from dock): 4 min

c. Other traffic entry and departure: 1.5 min

Off-loading (in minutes)	Number of Observations	Truck Arrivals (Between Times) (in minutes)	Number of Observations	Other Traffic (Between Times) (in minutes)	Number of Observations
50–60	12	0–20	28	0–5	16
60–70	36	20–40	39	5–10	34
70–80	21	40–60	22	10–15	25
80–90	17	60–80	8	15–20	15
90–100	14	80–100	3	20–25	10

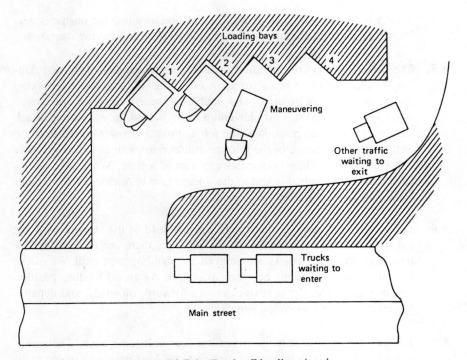

Figure P8.7.1 Truck off-loading site plan.

8.8. A three-worker labor crew is involved in trench excavation where one worker uses an air spade to loosen material and, by working two faces, keeps the other two workers occupied in shoveling material out of the trench. Every hour the spade operator is relieved by one of the shovelers so that in any 3-hr period each worker shovels for 2 hr and uses the air spade for 1 hr. If 15 minutes operation of the air spade generates 25 min of shoveling work, and the changeover of the work tasks takes 3 min, develop a capture mechanism model for the trench excavation operation.

8.9. Extend the mason problem described in Chapter 10 (Fig. 10.6) to include a mortar cycle. That is, the laborer on the ground also mixes mortar to be hoisted to the masons on the scaffold. Assume that the mason raises the hoist. The additional work tasks to be modeled should include mix mortar; load hoist with mortar; raise, unload, and lower hoist.

8.10. As a variation of the mason problem, consider the addition of a laborer on the scaffold with the masons. This laborer takes brick pallets from the hoist and either gives them to a mason or places them in storage on the scaffold. The scaffold laborers' priorities are to supply the mason directly from the hoist, to supply the mason from scaffold storage, and to replenish storage, in that order. Assume there is space on the scaffold for three brick pallets and that the laborer on the ground raises and lowers the hoist. Do not include the mortar cycle in this model.

8.11. Consider an extensive fast-moving pipeline construction project located in a remote area that requires camp facilities for construction labor. Assuming that the normal construction operation of easement clearance, excavation, pipe welding, pipe laying, backfill, and so forth are to be modeled as separate crew work tasks, develop a nonstationary process model for the pipeline construction, with travel to and from camp included in the labor workday. How would you develop a model to assist in the decision process for camp relocation as a factor in maximizing production?

8.12. Figure P8.12 is a schematic diagram of a roadway paving project. The diagram shows the stretch to be paved divided into five separate sections. The batch plant is located at the center of the stretch to be paved, and work progresses from left to right. The average travel time (*TT*) to sections 1 and 5 are the same. The loops indicate the haul cycle to these sections and *TT* 1 as the average duration of the haul cycle. The average travel time to sections 2 and 4 are given as *TT* 2 and the section 3 travel time is *TT* 3. The average travel times are

$$TT\ 1\ =\ 15\ \text{minutes}$$

$$TT\ 2\ =\ 10\ \text{minutes}$$

$$TT\ 3\ =\ 5\ \text{minutes}$$

Assume that 50 haul cycles are associated with each section and that work proceeds from left to right. Develop a CYCLONE model to switch the travel times as the work progresses.

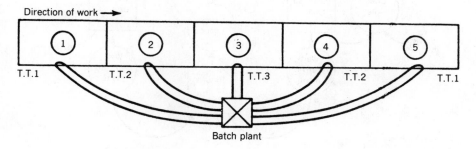

Figure P8.12 Schematic diagram of roadway paving project.

8.13. In Problem 8.12 above, how would you:

    a. Determine the variation in truck fleet size to ensure constant production at the paver?

    b. Determine the varying paver production for a constant fleet size?

    c. Establish the maximum haulage distance before batch plant relocation becomes necessary?

8.14. **a.** Given the model shown in Figure P8.14, for the transport of beam and deck sections to the site of a precast building project, develop a unit generation structure to insert deck and beam sections into the model and a capture mechanism to withdraw the trucks from the transport sequences for 16 hr during every 24-hr period.

    b. Modify the model to maintain the features required in part **a** above and provide for a break period of 5 min during each hour for each truck.

    c. Half of the time, deck and beam sections on trucks arriving to off-load can be lifted directly into position. Otherwise, the sections are off-loaded into a stockpile location. Thirty-three percent of the time, sections (if any) in stockpile are released for lift into the structure when the crane becomes idle because of a lack of trucks to be off-loaded. Modify the model of Problem 8.2 to reflect this feature of the crane's operation.

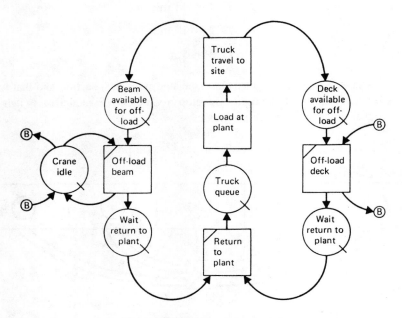

Figure P8.14 Transport model.

8.15. Using a probabilistic switching mechanism, develop a model for the operation of the worker hoist shown in Figure P8.15. Assume that the work has progressed to a stage such that the lifts demanded are as follows:

a. To zone I: 10%

b. To zone II: 20%

c. To zone III: 15%

d. To zone IV: 20%

e. To zone V: 35%

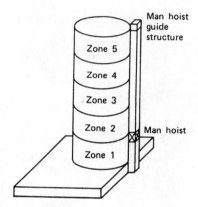

Figure P8.15 High-rise building with nonstationary process.

8.16. A river estuary is subject to tidal effects. Flat-bottom freight and passenger ferryboats can negotiate the river to city Alpha for unloading regardless of the tide level. Normal freighters can negotiate the river to and from the off-loading berths only in the interval from 2 hr before high tide to 2 hr after high tide. The docking berths have been deepened so that once a freighter has docked, it is not affected by the level of the tide. Because of crowded conditions, ships are required to wait at the mouth of the river until a berth becomes free. This is signaled from the unloading area by radio. Movement to and from the mouth of the estuary (to the unloading berths) requires the following times:

Flat-bottom freighter	25 min
Normal freighter	40 min
Passenger ferry	20 min

Five berths are available and can be used for off-loading either the ferry or the freighter. Because of narrowness of the channel, only one ship at a time can enter or leave. Priority on use of the channel is as follows:

Highest priority	Normal freighter
Next highest	Passenger ferry
Lowest	Flat-bottomed freighter

Arrival of both types of freighters at the mouth of the river is random. The ferry arrives every 120 min (± 20 min). The arrival patterns of the freight carriers have been observed as follows.

Interval Between Arrivals (in Hours)	Number of Observations, Freighter	Number of Observations, Flat-Bottomed Freighter
0–10	11	27
10–20	20	31
20–30	29	19
30–40	15	13
40–50	10	10
50–60	9	0
60–70	6	0

Berth time for the passenger ferry is fixed at 15 min. Unloading times for the freighters and flat-bottomed freight boats are as listed below.

Time (in Hours)	Freighter	Flat-Bottomed Freighter
0–1	0	22
1–2	0	33
2–3	17	25
3–4	33	14
4–5	19	6
5–6	18	0
6–7	13	0

The interval between high tides is 13 hr. Using a capture mechanism to handle the action of the tide, develop a CYCLONE model that describes the operation of this system. Run the CYCLONE model to determine the average time the normal freighters are delayed before entering to unload.

8.17. Using a green light system shown in Figure P8.17, suppose traffic is allowed to move from east to west for 3 min and then, by switching the light, traffic movement is allowed from west to east for 3 min. Suppose also that the system is to allow an extra half minute for any vehicle entering the construction just as the green light changes to transit before the green light in the opposite direction is switched on. The rate of entry of vehicles into the construction site will be assumed to be one every 5 sec when the green light is on. Design a mechanism to change the green light back and forth between the traffic lights located at the entry points to the construction site.

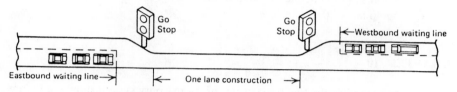

Figure P8.17 Haul road control with automatic light.

CHAPTER 9

9.1. Given the data shown in Table 3.1, calculate the balance point of the system production for systems of one to six trucks. Is the system working above or below the balance point?

9.2. The loader and arrival times for a system consisting of four trucks and two loaders are distributed as shown below.

Loader Time in Minutes		Back Cycle in Minutes	
Interval	Frequency	Interval	Frequency
0–0.5	2	4.0–4.5	4
0.51–1.0	5	4.51–5.0	10
1.01–1.5	10	5.01–5.5	28
1.51–2.0	20	5.51–6.0	32
2.01–2.5	25	6.51–7.0	20
2.51–3.0	25	7.01–7.5	4
3.01–3.5	10	7.51–8.0	2
3.51–4.0	2		100
4.01–4.5	1		
	100		

These tabulations represent field data on haul and load times. Calculate the production of this system deterministically and using the appropriate queueing nomograph in Appendix B. Develop a plot of the system production both deterministically and to include bunching for systems for 1 to 10 trucks.

9.3. If the system in Problem 9.2 (above) were working at the balance point, what production loss would result from bunching using the Morgan–Peterson approach?

9.4. Given the data in Table 9.1, use the linear congruential scheme (LCS) of Section 9.5 to generate 10 random bricklaying times. Set the value of the modulo $m = 20$.

CHAPTER 10

10.1. Hand-simulate the first three cycles of the masonry system shown in Figure 10.6, assuming that there are six stack locations on the scaffold and six masons supported by two laborers. Compare the productivity achieved with that found in the 29-min simulation described in this chapter.

10.2. Using hand simulation compare the impact in Problem 10.1 (above) of having all bricks stacked on the scaffold at the beginning of the shift versus having no bricks on the scaffold at the beginning of the operation. Plot the production transients for each system.

10.3. Several machines are repaired by a maintenance team when they break down. After repair the machines continue to run until they break down again. The repair times are given in Table P10.3.1. The running times are

TABLE P10.3.1 Repair Times

Interval (Hours)	Frequency
0 −0.4	1
0.41–0.8	5
0.81–1.2	17
1.21–1.6	37
1.61–2.00	61
2.01–2.40	78
2.41–2.80	62
2.81–3.20	93
3.21–3.60	42
3.61–4.00	28
4.01–4.40	36
4.41–4.80	27
4.81–5.20	13

TABLE P10.3.2 Running Times

Interval (Hours)	Frequency
0–2	4
2–4	19
4–6	47
6–8	79
8–10	76
10–12	67
12–14	54
14–16	39
16–18	34
18–20	26
20–22	20
22–24	15
24–26	10
26–28	7
28–30	3

given in Table P10.3.2. The machines are repaired in the sequence in which they break down; that is, the machine that breaks down first is repaired first. Do a hand simulation of five cycles of this system.

10.4. The CYCLONE model for a material hoist that lifts concrete blocks to the upper floors of a hospital project is shown in Figure P10.4. The hoist is loaded at the ground floor with a large fork lift and is unloaded at the work floor with a small hand-operated forklift. Five cycles of the small forklift are required to unload the material hoist. Conduct a hand simulation to determine the hourly production of this system. Productivity is measured at the COUNTER in 20-block increments. The element transit times are as follows:

Large Forklift Travel to Block Stockpile and Return Time in Minutes and Seconds		Small Forklift Travel to Stack	
3:30–3:45	2	0 –0:30	1
3:45–4:00	8	0:30–0:45	8
4:00–4:15	14	0:45–1:00	17
4:15–4:30	21	1:00–1:15	26
4:30–4:45	20	1:15–1:30	22
4:45–5:00	15	1:30–2:00	6
5:00–5:15	11		
5:15–5:30	9		

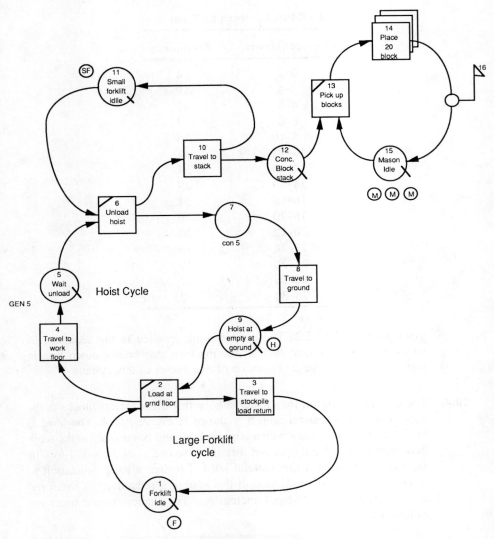

Figure P10.4 Material hoist model.

Hoist Load Time	Travel to Work Floor		Unload	Travel to Ground Floor	
Ground Floor					
Constant 1:30	0:15–0:20	4	Constant 1:00	0 –0:05	2
	0:20–0:25	10		0:05–0:10	9
	0:25–0:30	24		0:10–0:15	25
	0:30–0:35	26		0:15–0:20	28
	0:35–0:40	11		0:20–0:25	12
	0:40–0:45	5		0:25–0:30	4

Assume that pickup time for the mason is constant and equal to 1 min and that 3 masons are initialized at Queue Node "Mason Idle." Assume that the placement time per 20-block increment is twice the values given in Table 9.1.

Simulate two complete cycles of the hoist assuming the hoist is at ground level at $T = 0$. What is the project production of the system (blocks/hour) based on the two cycles which have been simulated?

CHAPTER 14

14.1. Using the activity times shown and initializing two trucks at the "Wt Load" position, estimate the idle time of the asphalt spreader in the model developed in Figure 11.5.

	Mean (Minutes)	Standard Deviation (Minutes)
Load at plant	5.0	0.75
Travel to job	30.0	5.00
Return to Load	25.0	5.00
Dump into spreader	7.0	2.00
Spread Asphalt	15.0	3.50
Reposition for pass	15.0	3.00
Compact asphalt section	20.0	5.00
Finish Section	30.0	4.00

Assume 50 "sections" of asphalt are placed.

14.2. Conduct a computer simulation of the precast concrete element plant discussed in this chapter to determine the idle time of resources at QUEUE node 18 ("Cure Position Free"). Assume that four crews are at QUEUE node 15 and eight forms are at QUEUE 24. All other initial flow unit information is shown in Figure 14.1. The times are as given in Table 14.1.

14.3. Although presented in this chapter in the context of a precast element plant operation, the model developed in Figure 14.1 can be used with appropriate time modification and task redesignation to model the casting operation of a steel plant. Assume that steel is poured in its molten state into a crucible in which it is allowed to cool to form an ingot. After initial cooling, each ingot is removed from the crucible and lifted into a cooling pit until it has reached the proper temperature for rolling. Following the cooling pit, the ingot is removed and rolled into the desired shape. A crew is involved in removing the ingot from the crucible after initial cooling and is also charged with cleaning the ingot for reuse. The parallel between the precast element plant operation and this process is fairly easy to detect. Modify the model of Figure 14.1 to handle the casting–rolling process just described and verify the system's operation using hand simulation. The times for the various operations are as shown below.

Operation	Time (in Minutes)
Pour ingot	10.0
Initial set	75.0
Lift into pit	20.0
Lift from pit	20.0
Cool in pit	200.0
Deslag (remove crucible)	15.0
Roll shape	50.0
Smelt	90.0
Clean Crucible	12.0

Determine the idle time for the crew and the crane in the process.

14.4. Consider the dry batch delivery and placement system discussed in Chapter 11 (see Figs. 11.1 and 11.2). Conduct a sensitivity analysis of this system by varying the number of trucks, crews, and mixers. Vary the number of trucks between 2 and 4, the number of crews from 1 to 2, and the number of mixers from 1 to 2. Which system is the most productive? If trucks cost $50 per hour, crews $100 per hour and mixers $200 per hour, which system is most cost-effective (e.g., lowest $/unit cost)?

14.5. Consider the earthmoving operation model shown on the next page.
  a. Determine the production per hour at the end of each cycle (i.e., the transient response) over a 20-cycle run for systems with one, two, three, and four trucks. All resources other than the trucks are fixed at one each. What is the balance point of this system in number of trucks.
  b. Assume that our objective is to maximize the utilization of the spreading dozer at node 14.
    (1) How would you approach the problem using simulation?
    (2) What combination of resources would you use to reach a maximum utilization of the dozer at node 14?
    (3) After soil is spread, it must be compacted. Assume that you have only one compactor. Modify the system model to include the com-

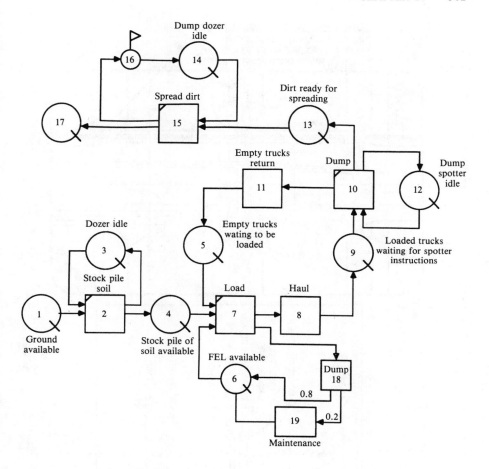

paction task. Assume that compacting one load requires 5.0 min. How does this impact system productivity?

14.6. Figure P14.6.1 represents the plan and profile drawing of a typical horizontal earth-boring project. The boring equipment and setup is illustrated in Figure P14.6.2. The method shown is the auger bore method. For more information on this method consult, the literature of the American Augers organization (American Augers, 1987). The following gives a brief description of the process. All information in this problem is based on the Doctoral thesis of David T. Iseley (1989). A horizontal earth-boring (HEB) contractor normally works as a subcontractor on projects of this nature. The bid price is based on being able to move in, execute the bore, then move to the next project with minimal delays. When delays occur, they are normally significant. The contract usually provides some way of compensating the subcontractor on an hourly basis when a delay takes place

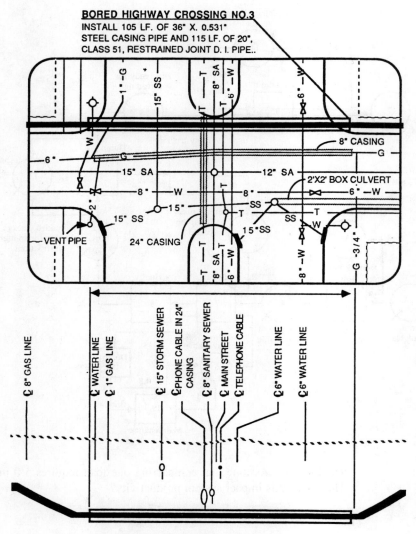

BORED HIGHWAY CROSSING NO.3
INSTALL 105 LF. OF 36" X. 0.531"
STEEL CASING PIPE AND 115 LF. OF 20",
CLASS 51, RESTRAINED JOINT D. I. PIPE..

Figure P14.6   Plan and profile view.

via "a changed condition setup" clause. The most common forms of changed conditions in "trenchless excavation" are due to subsurface conditions. If, for example, the contractor based his bid on dry stable clay in the absence of other information, and he encounters wet unstable clay at the job site, this will adversely impact the duration of major work tasks. This will lengthen the project duration and delay the completion date. A typical HEB project is composed of the following work tasks in the following specified order:

(A) The boring machine track is set in the prepared boring pit and adjusted for line and grade.

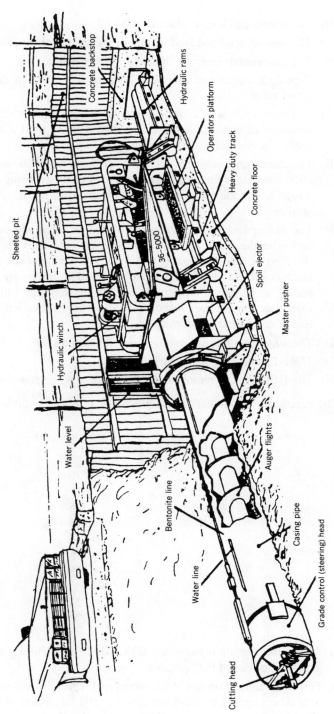

Concrete backstop

Hydraulic rams

Operators platform

Heavy duty track

Concrete floor

Spoil ejector

Master pusher

36-5000

Sheeted pit

Hydraulic winch

Water level

Auger flights

Bentonite line

Casing pipe

Water line

Grade control (steering) head

Cutting head

Figure P14.6.2 Track-type auger horizontal earth boring.

(B) The boring machine is placed on the tracks.

(C) The cutting head and leading end of the auger are installed.

(D) The casing and auger are attached to the machine.

(E) Pushing is started and continued for a full 4-ft stroke of the hydraulic thrust rams.

(F) The push cycle ends and the hydraulic thrust rams are retracted. The sequence of tasks D and E is repeated 5 times until a 20-ft casing section is inserted.

(G) The casing and auger are disconnected from the HEB machine. The HEB machine is then retracted to the rear of the pit using a hydraulic winch and made ready to receive the next 20-ft casing and auger section.

(H) The casing and auger section are placed on track and the auger attached at the leading end. The casing is welded to the previously installed casing. The casing and auger are then attached to the HEB machine. (Note that the crane will be freed at this point). Pushing is started again (work tasks 5, 6, and 7) and repeated until the 20-ft casing section has been inserted. This step is repeated until 100 linear feet of casing has been installed.

(I) The auger sections are removed one flight at a time.

(J) The HEB machine is finally removed.

(K) The HEB track is removed and the process is completed.

The durations of each activity are given in the following table:

Activity Label	Duration (min)
A	50
B	20
C	17
D	25
E	18
F	6
G	20
H	35
I	20
J	25
K	30

a. Model the operation using the CYCLONE methodology and draw a graphical model of the operation.

b. Perform a simulation run of the model and report the following: (1) the total time required to finish the process (100 linear feet of casing) and (2) the percent time that the HEB machine is idle.

c. Once the model is running, you are to use it to substantiate how much money the subcontractor should get given that (1) the agreement indicated that soil borings showed stable dry clay and should the contractor be delayed the compensation rate would be $200 per hour, (2) the subcontractor can show that the durations given in the table are representative of the process based on historical data, and (3) the subcontractor showed that due to encountering wet unstable clay task, "E" (above) required 80 min instead of the estimated 18 min.

d. To make the model more realistic, we assume that for 80% of the time the clay encountered was dry and stable. Only in for 20% of the time was the clay wet and unstable. Develop an enhanced model to account for this situation.

14.7. This problem involves the modeling and simulation of a tunneling operation. In particular, the earth pressure balance (EPB) shield method will be considered. The process of tunneling with the EPB machine starts with digging and preparation of a pit to form a shaft. When the shaft is prepared the machine is assembled at the bottom of the shaft. When the machine advances, muck has to be removed from the tunnel, precast concrete units have to be brought in to the excavation face, the liners are then placed to support the tunnel, the supported face is then grouted, a jacking system—used by the EPB machine to advance—is then placed on the liners, and the machine adjusted and advanced. The liners and the muck cars are carried in and out of the tunnel using a train traveling on track. After the placement of the four rings the track has to be extended. The machine advances in increments of four feet, the amount necessary to install one ring of liners.

Objectives of the Simulation

The model of this process has to be prepared in such a way as to enable you to determine the (1) attainable levels of production, (2) the time required to complete the first 200 ft of the tunnel, and (3) what, if anything, can be done to increase the advance rate of the tunnel.

Description of the Process

The process can be divided into two basic parts: (1) initial excavation of the shaft and placement of the EPB machine and (2) excavation once the tunneling operation is in progress. The following describes each of the two main parts of the process:

1. Initial excavation. This involves the following tasks in the sequence presented:
   a. Unload the EPB machine parts from trucks and preassemble the main components; assume it takes 2 days. No prerequisites.
   b. Excavate the shaft and prepare the bottom to receive the machine; assume 2 days. No prerequisites.

    c. Set up the machine and install the thrust reactor. Assume 1 day. Prerequisites are a and b.

    d. Hook electrical power to EPB; assume 2 hr.

    e. Push for the initial 4 ft; assume 20 min.

    f. Dump muck into muck cars and remove cars from the shaft to the surface; assume 1 hr 30 min.

    g. Place the first set of liners; assume 45 min.

2. Excavation after initial phase is completed.

    a. Place jacks and the jacking system (3 min) requires that the EPB and its crew be available, and the placement of liners and grouting tasks completed.

    b. When the muck cars are properly placed, start excavation for 4 ft (15 min).

    c. After the 4-ft excavation perform routine maintenance on machine (5 min).

    d. The muck cars have to be removed from the tunnel and emptied outside. If the track is not being used to bring in liners this task can be carried out. Traveling takes about 5 min. Once in the shaft area, a hoist transports the cars outside the tunnel and back to the bottom of the shaft. When emptied, this takes about 15 min, at which point the excavation can be reinitiated, i.e., task "a" and the sequence of following tasks repeated (if the liners have been already placed). The cars usually return in about 5 min; traveling back to the excavation face takes an additional 5 min. (Again note that the hoist will transport both the liners and the muck cars, so each has to wait until the hoist is free.)

    e. The jacks and jacking system are removed; it takes about 3 min to do that. Once the jacks are removed, the liners are placed; this requires the liners to be available at the excavation face location (the liners are carried by the train from shaft location to excavation face—described later). It takes the crew about 30 min to place one full ring (seven liners) on the average. After placing the liners, the crew performs the grouting task, which takes about 15 min. The train, meanwhile, would have returned to the shaft (if the track was not being used to transport the muck cars); it takes about 5 min to travel from the location of the excavation face to the shaft location. A new set of liners (seven of them) would then be lowered to the shaft location using the hoist (this takes 15 min). Once on the train, they are transported to the excavation face (5 min); this enables a new sequence of task "e" to be reinitiated. It is to be noted, at this point, that at the completion of four rings, the track is extended (this takes 30 min). The track cannot be used until this "extension" task is completely done.

Required Enhancements of the Original Model

Provide an enhancement to the system to model the breakdown of the EPB machine. You are given that the machine was down once during the observation of thirty 4-ft pushes. It takes anywhere between 2 and 4 hr to fix the machine depending on the problem encountered. How do the new results compare to the original model with regard to the tunnel advance rate. Provide an enhancement that would model the possibility of hitting a boulder while excavating. You are told that there is a chance of 2 in 100 that a boulder might be hit. Excavating the boulder and resuming the operation takes anywhere between 60 and 360 min depending on its size, location, and properties. How do the new results compare to the original model with regard to the tunnel advance rate? Simulate the enhanced model 30 times and provide an average for the tunnel advance rate based on the results. How do the new results compare to the original deterministic model with regard to the tunnel advance rate?

CHAPTER 15

15.1. Create a CYCLONE network from the schedule shown in Figure P15.1.1. Assume that the hydrotest activity does not pass 15% of the time and the piping must be reworked. After rework, the pipes are again hydrotested. Likewise, assume that the system test does not pass 5% of the time. In this event, the entire system must be inspected and reworked, and the system again checked. Include these rework loops in your CYCLONE network.

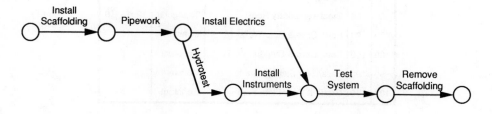

	Type	Duration (Low,Mode,High)
Install Scaffolding	Constant	5
Pipework	Triangular	18,20,24
Hydrotest	Constant	3
Install Electrics	Triangular	12,15,20
Install Instruments	Triangular	9,10,14
Test System	Constant	5
Remove Scaffolding	Constant	3
Rework Pipe	Triangular	1, 2, 3
Rework System	Triangular	1, 3, 5

Figure P15.1.1 Retrofit schedule.

15.2. Draw a CPM diagram using the activity relationships shown below. Convert this diagram into a CYCLONE network and run it. The activity durations are assumed to be represented by triangular distributions. Determine the expected time of completion and the probable critical path.

i	j	Activity	Type	Duration
10	20	Prefab Wall Forms	Constant	2
10	30	Excavate Cols and Walls	Constant	3
10	40	Let Elec and Mech Subcontract	Triangular	3, 4, 8
20	60	Deliver Wall Forms	Constant	4
30	40	Dummy	-----	0
30	60	Forms, Pour & Cure Wall & Col Ftg	Triangular	6, 7, 8
40	50	Rough-In Plumbing	Triangular	5, 7, 10
40	70	Install Conduit	Triangular	9, 11, 15
50	70	Dummy	-----	0
60	70	Erect Wall FOrms & Steel	Constant	9
60	80	Fabricate & Set Interior Column Forms	Constant	6
60	100	Erect Temporary Roof	Triangular	12, 16, 18
70	90	Pour, Cure & Strip Walls	Constant	10
80	100	Pour, Cure & Strip Int. Walls	Constant	6
90	100	Backfill for Slab on Grade	Constant	1
100	110	Grade & Pour Floor Slab	Constant	5

15.3. Consider the following decision process involving the launch of an orbiting satellite. If the launch countdown is unsuccessful because of mechanical, electrical, or fuel problems, the launch is canceled. The launch vehicle is then inspected, repaired, and made ready for the second countdown. If the second countdown is unsuccessful, the vehicle is again inspected, repaired, and made ready for a third launch. Should the third launch be unsuccessful, the mission is canceled and the vehicle is sent back to the manufacturer. Construct a CYCLONE network to simulate this decision process using the information given in Figure P15.3.

15.4. A contractor is considering a decision to open a branch office in another state. Resources are available to open either a large office or a small office. The costs associated with each decision are shown in millions of dollars on the decision tree in Fig. P15.4. The contractor's assessment of the probable business climate and associated revenue are also shown on the decision tree. Convert the decision tree to a CYCLONE diagram and simulate the process 100 times to determine the best strategy.

Task	Probability		Cost
	OK	Not OK	
Launch Countdown 1	0.80	0.20	$1.20M
First Rework			$0.20M
Launch Countdown 1	0.90	0.10	$0.60M
Second Rework			$0.15M
Launch Countdown 1	0.98	0.02	$0.60M
Cancel			$0.85M

Figure P15.3 Launch information.

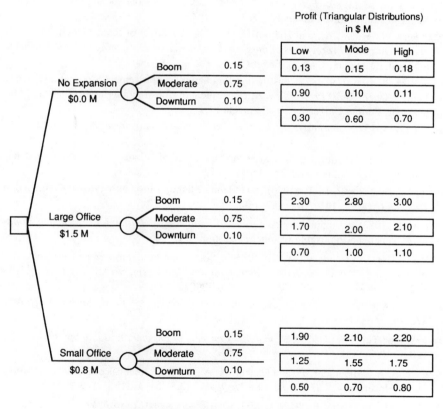

Figure P15.4 Contractor's decision tree.

BIBLIOGRAPHY

AbouRizk, S. M. (1985), "Sensitivity Analysis of Construction Operations," M.S. Special Topic, School of Civil Engineering, Georgia Institute of Technology.

AbouRizk, S. M. (1988), "Statistical Considerations in Simulating Construction Operations," *Independent Research Study Report*, Division of Construction Engineering and Management, Purdue University, W. Lafayette, IN.

AbouRizk, S. M. (1989), *BetaFit User's Guide*, Division of Construction Engineering and Management, Purdue University.

AbouRizk, S., Hijazi, A., and Halpin, D. W. (1989), "Effect of Input Modeling on Simulation Output," *Proceedings of the Sixth Conference on Computing in Civil Engineering*, ASCE, Atlanta, GA, pp. 605–612.

AbouRizk, S. et al. (1988), "Construction Productivity Improvement Study of a Concrete Pouring Operation," *Project Report*, Department of Construction Engineering and Management, Purdue University, W. Lafayette, IN.

Adrian, J. (1973), *Quantitative Methods in Construction Management*, American Elsevier, New York.

Adrian, J. J. (1974), thesis presented to the University of Illinois, Urbana, in partial fulfillment of the requirements for a Ph.D.

Adrian, J., and Boyer, L. T. (1976), "Modeling Method—Productivity," *J. Construction Division* (ASCE), **102**(C01), 157–168.

Adrian, J. (1983), "Productivity Analysis and Estimating," *Building Construction Handbook*, American Publishing Company, Ruston, VA.

Ahuja, N. H., and Nandakumar, V. (1985), "Simulation Model to Forecast Project Completion Time." *J. Construction Engineering and Management* (ASCE), **111**(4), 325–342.

American Augers, Inc., (1987) *Technical Manual-Horizontal Earth Boring Machines*, Wooster, OH, Feb.

Ang, A. and Tang, W. (1975), *Probability Concepts in Engineering Planning and Design*, Wiley, New York.

Ang, H.-S. A., Abdelnour, J., and Chaker, A. A. (1975), "Analysis of Activity Networks Under Uncertainty," *J. Engineering Mechanics* (ASCE), **101**(EM4), 373–387.

Antill, J. M., and Ryan, P. W. S. (1967), *Civil Engineering Construction*, 3rd ed., Angus and Robertson, London.

Antill, J. M., and Woodhead, R. W. (1985), *Critical Path Methods in Construction Practice*, Wiley, New York.

"Arch Halves Fall, Making Pivotal Link in German," (1985), *Engineering News-Record*, June 13, pp. 30–34.

Ashley, D. B. (1980), "Simulation of Repetitive-Unit Construction." *J. Construction Division* (ASCE), **106**(C02), 185–194.

Banks, J., and Carons, J. S. (1984), *Discrete-Event System Simulation*, Prentice-Hall, Englewood Cliffs, NJ.

Beckman, R. J., and Tietjen, G. L. (1978), "Maximum Likelihood Estimation for the Beta Distribution," *J. Statist. Comput. Simul.*, **7**, 253–258.

Benjamin, N., and Greenwald, T. (1973), "Simulating Effect of Weather on Construction," *J. Construction Division* (ASCE), **99**(C01), 175–190.

Bernold, L. E., and Halpin, D. W. (1984), "Microcomputer Cost Optimization of Earth Moving Operation," CIB W-65, *Proceedings of the 4th International Symposium on Organization and Management of Construction*, Vol. 11, Waterloo, Ontario, Canada, pp. 333–341.

Bernold, L. E. (1985), "Productivity Transients in Construction Processes," Ph.D. Thesis, Georgia Institute of Technology, Atlanta, GA.

Berrios, J., and Halpin, D. W. (1988), "Construction Optimization Using Resource Simulation," *Concrete International—Design and Construction*, **10**(6), 38–42.

Biles, W. E. (1987), "Introduction to Simulation," *Proceedings of the 1987 Winter Simulation Conference*, IEEE, pp. 7–15.

Borcherding, J. D. (1977), "Cost Control Simulation and Decision Making," *J. Construction Division* (ASCE), **103**(C04), 577–591.

Bradshaw, L. M., Jr. (1978), "A Critical Review and Analysis of the Method Productivity Delay Model," M.S. Special Topic, School of Civil Engineering, Georgia Institute of Technology.

Brately, P., Fox, B., and Schrage, L. (1983), *A Guide to Simulation*, Springer-Verlag, New York.

Brooks, A. C., and L. R. Schaffer (undated), "Queueing model for Production Forecasts of Construction Operations," unpublished report, Department of Civil Engineering, University of Illinois, Urbana.

Carr, R. I., and Meyer, Walter L. (1974), "Planning Construction of Repetitive Building Units," *J. Construction Division* (ASCE), **100**(C03), 403–412.

Carr, R. I. (1979), "Simulation of Construction Project Duration," *J. Construction Division* (ASCE), **105**(C02), 117–128.

Carroll, J. M. (1987), *Simulation Using Personal Computers*, Prentice-Hall, Englewood Cliffs, NJ.

Caterpillar Performance Handbook (1987), 18th ed., Caterpillar Tractor Company, Peoria, IL, October.

Chang, D. Y., and Carr, R. I. (1987), "RESQUE: A Resource Oriented Simulation System for Multiple Resource Constrained Processes," *Proceedings of the PMI Seminar/Symposium*, Milwaukee, WI, pp. 4–19.

Chang, T. C., and Wysk, R. A. (1983), "CAD/Generative Process Planning with TIPPS," *J. Manufacturing Systems*, **2**(2), 127–135.

Chrazanwski, E. N., and Johnston, D. (1986), "Application of Linear Scheduling," *J. Construction Engineering and Management* (ASCE), **112**(4), 476–491.

Clark, C. E. (1962). "The PERT Model for the Distribution of an Activity Time," *Operations Research*, **10**(3), 405–406.

Clemmens, J. P., and Willenbrock, J. H. (1978), "The SCRAPESIM Computer Simulation," *J. Construction Division* (ASCE), **104**(C04), 419–435.

Coon, Helen (1964), "Note on William A. Donaldson's "The Estimation of the Mean and Variance of a PERT Activity Time," *Operations Research*, **13**, 386–387.

Cottrell, G. B. (1989), *Class Report: Productivity Study of Indoor Training Facility*, Division of Construction Engineering and Management, Purdue University, W. Lafayette, IN.

Crandall, K. C. (1976), "Probabilistic Time Scheduling," *J. Construction Division* (ASCE), **102**(CO3), 415–423.

Crandall, K., and Woolery, J. (1982), "Schedule Development under Stochastic Scheduling," *J. Construction Engineering and Management* (ASCE), **108**(C02), 321–329.

Dabbas, M. A. A. (1981), "Computerized Decision Making in Construction," Ph.D. thesis, Georgia Institute of Technology, Atlanta, GA.

Dabbas, M. A. A., and Halpin, D. W. (1982), "Integrated Project and Process Management," *J. Construction Division* (ASCE), **108**(C03), 361–374.

Debrota, D., Roberts, S. D., Dittus, R. S., and Wilson, J. R. (1988), "Visual Interactive Fitting of Probability Distributions," *Simulation*, **52**, 199–205.

Depew, D., Lutz, J., and Nekoomaram, S. (1986), "Construction Methods Improvement Study," unpublished student report, Purdue University, W. Lafayette, IN.

Digman, L. A. (1967), "PERT/LOB: Life-Cycle Technique," *J. Industrial Engineering*, Vol. 18, No. 2, pp. 154–158.

Donaldson, W. A. (1965), "The Estimation of the Mean and Variance of a PERT Activity Time," *Operations Research*, **13**, 382–385.

Douglas, D. E. (1978), "PERT and Simulation," *Winter Simulation Conference Proceedings*, IEEE, Vol. 1, pp. 88–98.

Dressler, J. (1974), "Stochastic Scheduling of Linear Construction Sites," *J. Construction Division* (ASCE), 571–588, Vol. 100, No. C04.

El Sheikh, Asim, A. R., Paul, R. J., Harding, A. S., and Balmer, D. W. (1987), "A Microcomputer-Based Simulation Study of a Port," *J. Operational Research Society*, **38**(8), 673–681.

Eppen, G. D., and Gould, F. J. (1985), *Quantitative Concepts for Management Decision Making Without Algorithms*, Prentice-Hall, Englewood Cliffs, NJ.

Fechtig, R. (1986), "Renewal of the Old Quaibridge in Zurich," *International Association for Bridge and Structural Engineering Journal*, J-30/86, May.

Feller, W. (1957), *An Introduction to Probability Theory and Its Applications*, Vol. 1, 2nd ed., Wiley, New York.

Fishman, G. S. (1973), *Concepts and Methods in Discrete Event Digital Simulation*, Wiley, New York.

Gaarslev, A. (1969), "Stochastic Models to Estimate the Production of Material Handling Systems in the Construction Industry," *Technical Report No. 111*, The Construction Institute, Stanford University, Palo Alto, CA, August.

Gates, M., and Scarpa, A. (1978), "Optimum Number of Crews," *J. Construction Engineering and Management* (ASCE) **104**, pp. 123–132.

Gordon, G. (1978), *System Simulation*, Prentice-Hall, Englewood Cliffs, NJ.

Griffis, F. H., Jr. (1968), "Optimizing Haul Fleet Size Using Queueing Theory," *J. Construction Division* (ASCE), **94**(C01), 531–544.

Grubbs, F. E. (1962), "Attempts to Validate Certain PERT Statistics or 'Picking on PERT'," *Operations Research*, **10**, 912–915.

Haas, R. and Halpin, D. W. (1978), "Production Modeling of Precast Transportation Structures," *Proceedings of International Congress on the Application of Mathematics in Engineering*, Weimar, Germany.

Hahn, G. J., and Shapiro, S. S. (1967), *Statistical Models in Engineering*, Wiley, New York.

Halpin, D. W., and Happ, W. W. (1971), "Digital Simulation of Equipment Allocation for Corps of Engineer Construction Planning," *Proceedings of the Seventeenth Conference of Design of Experiments in Army Research*, Washington, DC.

Halpin, D. W., and Happ, W. W. (1972), "Network Simulation of Construction Operations," *Proceedings, Third International Congress of Project Planning by Network Techniques*, Stockholm, Sweden.

Halpin, D. W., and Woodhead, R. W. (1972), "A Network-Based Methodology for the Management Modeling of Complex Projects," *Proceedings, Third International Congress on Project Planning by Network Techniques*, Stockholm, Sweden.

Halpin, D. W. and Woodhead, R. W. (1973), *CONSTRUCTO—A Heuristic Game for Construction Management*, University of Illinois Press, Urbana.

Halpin, D. W. (1973), "An Investigation of the Use of Simulation Networks for Modeling Construction Operations," Ph.D. thesis, University of Illinois, Urbana-Champaign.

Halpin, D. W., and Woodhead, R. W. (1976), *Design of Construction and Process Operations*, Wiley, New York.

Halpin, D. W. (1977), "CYCLONE—A Method for Modeling Job Site Processes," *J. Construction Division* (ASCE), **103**(C03), 489–499.

Halpin, D. W. (1990), *MicroCYCLONE User's Manual*, Learning Systems Inc., W. Lafayette, IN.

Halpin, D. W., and McCahill, D. (1986), "Modeling Construction Operations in the Classroom," *Proceedings, 4th Conference on Computing in Civil Engineering*, Boston, MA, October.

Halpin, D. W. (1987), "Construction Optimization Using Micro Computers," *Proceedings of CIB W-65 Symposium 1987*, London, September, pp. 488–499.

Halpin, D. W., Hijazi, A., and AbouRizk, S. (1989), "Planning of Non-Stationary Construction Processes," *Proceedings, 7th National Conference on Microcomputers in Civil Engineering*, November 8–9, Orlando, FL.

Halpin, Daniel W., AbouRizk, S., and Hijazi, A. (1989), "Sensitivity Analysis of Construction Operations," *Proceedings, 7th National Conference on Microcomputers in Civil Engineering*, November 8–9, Orlando, FL.

Handa, V. K., and Barcia, R. M., (1986), "Linear Scheduling Using Optimal Control Theory," *J. Construction Division* (ASCE), **112**(3), 387–393.

Handa, V. K., and Barcia, R. M. (1986), "Construction Production Planning," *J. Construction Engineering and Management* (ASCE), **112**(2), 163–177.

Harris, F. C., and Evans, J. B. (1977), "Road Construction: Simulation Game for Site Managers, *J. Construction Division* (ASCE), **103**(C03), 407–414.

Harris, R. B. (1978), *Precedence and Arrow Networking Techniques for Construction*, Wiley, New York.

Havers, J. A., and Stubbs, F. W., Jr. (eds.) (1971), *Handbook of Heavy Construction*, 2nd ed., McGraw-Hill, New York.

Hijazi, A. M. (1989), "Simulation Analysis of Linear Construction Processes," Ph.D. dissertation, School of Civil Engineering, Purdue University, W. Lafayette, IN, May 1989.

Hill, S. L. (1989), "Analysis of Construction Input Variables on Simulation Processes," M.S. Special Topic, Division of Construction Engineering and Management, Purdue University, Lafayette, IN, December 1989.

Howard, R. A. (1960), *Dynamic Programming and Markov Processes*, Technology Press of Massachusetts Institute of Technology, Cambridge, MA.

Iseley, D. T. (1988), "Automated Methods for the Trenchless Placement of Underground Utility Systems," Ph.D. dissertation, School of Civil Engineering, Purdue University, W. Lafayette, IN.

Jackson, J. R. (1957), "Networks of Waiting Lines," *Operations Research* **5**(4), 518–521.

Johnson, N. L. (1949), "System Soft Frequency Curves Generated by Methods of Translation," *Biometrika*, **36,** 149–176.

Johnston, D. W. (1981), "Linear Scheduling Method for Highway Construction," *J. Construction Division* (ASCE), **107**(C02), 274–281.

Kalk, A. (1980), "INSIGHT–Interactive Simulation of Construction Operations Using Graphical Techniques," *Technical Report No. 238*, Construction Institute, Department of Civil Engineering, Stanford University, Stanford, CA.

Kelton, W. D. (1986), "Statistical Design and Analysis," *Proceedings, 1986 Winter Simulation Conference IEE*, Washington, DC, pp. 45–51.

Key, J. M. (1987), "Earthmoving and Heavy Equipment," *J. Construction Engineering and Management* (ASCE), **113**(4), 611–622.

Khisty, C. J. (1970), "The Application of the Line of Balance Technique to the Construction Industry," *Indian Concrete J.*, (July).

Knot, J. L., and Woodhead, R. W. (1980), "The CYCLONE-Timelapse Analysis System," *Construction Monographs*, School of Civil Engineering, University of New South Wales, Australia.

Law, A. M., and Kelton, D. W. (1982), *Simulation Modeling and Analysis*, McGraw Hill, New York.

Law, A. M., and Vincent, S. G. (1985), *UNIFIT User's Guide*, Simulation Modeling Analysis Company, Tucson, AZ.

Law, A. M., and McComas, M. G. (1986), "Pitfalls in the Simulation of Manufacturing Systems," *Proceedings, 1986 Winter Simulation Conference*, Washington, DC, pp. 539–542.

Law, A. M., and Vincent, S. G. (1986), "A Tutorial on UNIFIT," *Proceedings, 1986 Winter Simulation Conference*," IEE, Washington, DC, pp. 218–222.

Levitt, H. P. (1968), "Computerized Line of Balance," *J. Industrial Engineering*, **19**(2), 61–66.

Livingston, E., and Acenbrak, S. (1987), "Interaction of Concrete and Placement with Crew Size and Productivity," unpublished student report, University of Maryland, College Park, MD.

Lluch, J. F. (1981), "Analysis of Construction Operations Using Microcomputers," Ph.D. thesis, Georgia Institute of Technology, Atlanta, GA.

Lluch, J., and Halpin, D. W. (1982), "Construction Operations and Microcomputers," *J. Construction Division* (ASCE), **108**(C01), 129–145.

Lumsden, P. (1968), *The Line of Balance Method*, Pergamon Press, London, U.K.

MacCrimmon, K. R., and Ryavec, C. A. (1964), "An Analytical Study of the PERT Assumptions," *Operations Research*, **12**(1), 16–37.

Martin, G. C. (undated), "Hopper-Truck Queueing Theory Model," unpublished report, Department of Civil Engineering, University of Illinois, Urbana.

Mayer, R. H., and Stark, R. (1981), "Earthmoving Logistics," *J. Construction Division* (ASCE), **107**(C02), 297–312.

McBridge, W., and McClelland, C. (1967), "PERT and the Beta Distribution," *IEEE Transactions on Engineering Management*, **EM-14**(4), 166–169.

MicroCYCLONE User's Manual, Version 2.0, May 1989, School of Civil Engineering, Purdue University, W. Lafayette, IN, May 1989.

Mihram, G. A. (1972), *Simulation: Statistical Foundations and Methodology*, Academic Press, New York.

Mitrani, I. (1982), *Simulation Techniques for Discrete Event Systems*, Cambridge University Press, Cambridge, U.K.

Mohieldin, Y. A. (1989), "Analysis of Construction Processes with Non-Stationary Work Task Durations," Ph.D. dissertation, Civil Engineering Dept., University of Maryland, College Park, MD, May 1989.

Morgan, W. C., and Peterson, L. (1968), "Determining Shovel–Truck Productivity," *Mining Engineering*, December, pp 76–80.

Naaman, M. A. (1974), "Networking Methods for Project Planning and Control," *J. Construction Division* (ASCE), 357–372.

Nandgaonkar, S. M. (1981), "Earthwork Transportation Allocations: Operations Research," *J. Construction Division* (ASCE), **107**(C02), 373–392.

Naylor, T. H., Balintfy, J. L., Burdick, D. S., and Chu, K. (1966), *Computer Simulation Techniques*, Wiley, New York.

Neelamkavil, F. (1987), *Computer Simulation and Modelling*, Wiley, Chichester, U.K.

Neter, J., Wasserman, W., and Whitmore, G. (1988), *Applied Statistics*, 3rd ed., Allyn and Bacon, Boston.

Niederhauser, D. (1984), "The Application of Standard Models in MicroCYCLONE," M.S. Special Topic, School of Civil Engineering, Georgia Institute of Technology, Atlanta.

Nunnally, S. (1980), Simulation of Construction Operations, *ASEE Annual Conference Proceedings*, Engineering Education for the 21st Century, Vol. 2, pp. 411–418.

Nunnally, S. W. (1987), *Construction Methods and Management*, Prentice-Hall, Englewood Cliffs, NJ.

O'Brien, J. J. (1975), "VPM Scheduling for High Rise Buildings," *J. Construction Division* (ASCE), **101**(C04), 895–905.

O'Brien, J. J., Kreitzberg, F. C., and Mikes, W. F. (1985), "Network Scheduling Variations for Repetitive Work," *J. Construction Engineering and Management* (ASCE), **111**(2), 105–116.

O'Connor, J. T. (1985), "Impact of Constructability Improvement," *J. Construction Engineering and Management* (ASCE), December, 404–410.

O'Neill, R. R. (1956), "An Engineering Analysis of Cargo Handling vs. Simulation of Cargo Handling Systems," *Report 56-37*, Department of Engineering, University of California, Los Angeles, September.

O'Shea, J. B., Slutkin, G. N., and Shaffer, L. R. (1964), "An Application of the Theory of Queues to the Forecasting of Shovel–Truck Fleet Productions," *Construction Research Series No. 3*, Department of Civil Engineering, University of Illinois, Urbana, June.

Palm, C. (1947), "The Distribution of Repairmen in Servicing Automatic Machines," *Industritidningen Norden*, **75**, 75–80.

Parker, H. W., and Oglesby, C. H. (1972), *Methods Improvement for Construction Managers*, McGraw-Hill, New York.

Perera, S. (1983), "Resource Sharing in Linear Construction," *J. Construction Engineering and Management* (ASCE), **109**(1), 102–111.

Paul, R. J., and Chew, S. T. (1987), "Simulation Modelling Using an Interactive Simulation Program Generators," *J. Operational Research Society*, **38**(8), 735–752.

Paulson, B. C. (1978), "Interactive Graphics for Simulating Construction Operations," *J. Construction Division* (ASCE), **104**(C01), 69–76.

Paulson, B. C., Jr., Douglas, S. A., Kalk, A., Touran, A., and Victor, G. A., (1983), "Simulation and Analysis of Construction Operations," *J. Technical Topics in Civil Engineering* (ASCE), August 1983, Vol. 109, No. 2, pp. 89–104.

Paulson, B. C., Chan, W. T., and Koo, C. C. (1987), "Construction Operations Simulation by Microcomputer," *J. Construction Engineering and Management* (ASCE), **113**(2), 301–314.

Pearson, E. S., D'Agostino, R. B., and Bowman, K. O. (1977), "Test for Departure from Normality: Comparison of Powers," *Biometrica*, **64**, 231–246.

Peer, S. (1974), "Network Analysis and Construction Planning and Control," *J. Construction Division* (ASCE), 203–210.

Peurifoy, R. L., and Ledbetter, W. B. (1985), *Construction Planning, Equipment and Methods*, 2nd ed., McGraw-Hill, New York.

Pilcher, R., and Flood, I. (1984), "The Use of Simulation Models in Construction," *Proceedings of the Institution of Civil Engineers*, Part 1, 76, London, U.K., pp. 635–652.

Poole, T. G., and Szymankiewicz, J. Z. (1977), *Using Simulation to Solve Problems*, McGraw-Hill, London.

Pritsker, A. A. B., and Happ, W. W. (1966), "GERTS: Part I—Fundamentals," *J. Industrial Engineering*, **17**(5), 267–274.

Pritsker, A. A. B., and Whitehouse, G. W. (1966), "GERT: Part II—Probabilistic and Industrial Engineering Applications," *J. Industrial Engineering*, **17**(6), 293–301.

Pritsker, A. A. B., and Kiviat, P. J. (1969), *Simulation with GASP II*, Prentice-Hall, Englewood Cliffs, NJ.

Pritsker, A. A. B. (1985), *Introduction to Simulation and SLAM-II*, Wiley, New York.

Rainer, R. K. (1968), "Predicting Productivity of One or Two Elevators for Construction of High-Rise Buildings," Ph.D. dissertation, Auburn University, Auburn, AL.

RAND Corporation (1955), *A Million Random Digits with 100,000 Normal Deviates*, The Free Press, Glencoe, IL.

Riggs, L. S. (1979), "Sensitivity Analysis of Construction Operations," Ph.D. thesis, Georgia Institute of Technology, Atlanta.

Riggs, L. S. (1980), "Sensitivity Analysis of Construction Operations," *J. Technical Councils* (ASCE), **106**, 145–163.

Riggs, L. S. (1980), "Simulation of Construction Operations," *J. Technical Councils* (ASCE), **105**(TC1), 145–163.

Riggs, L. S. (1989), "Risk Management in CPM Networks," Microcomputers in Civil Engineering, Elsevier Applied Science, Vol. 3, No. 3, Sept., pp. 229–235.

Rincones, L. (1984), "Computer Graphics Representation of Standard CYCLONE Models," M.S. Special Topic, School of Civil Engineering, Georgia Institute of Technology, Atlanta.

Ringwald, R. C. (1987), "Bunching Theory Applied to Minimize Cost," *J. Construction Division* (ASCE), **113**(2), 321–326.

Rounds, J. L., Hendrick, D., and Higgins, S. (1985), "Project Management Simulation Training Game," *J. Management in Engineering* (ASCE), **2**(4), 272–279.

Royston, J. P. (1982), "The W Test for Normality," *Applied Statistics*, Royal Statistical Society, London, pp. 176–224.

Russell, A., and Caselton, W. (1988), "Extensions to Linear Scheduling Optimization," *J. Construction Engineering and Management* (ASCE), 36–52.

Selinger, S. (1980), "Construction Planning for Linear Projects," *J. Construction Division* (ASCE), 195–205.

Shahabadi, A. (1985), "Sectional Method of Construction," *Betonwerk & Fertigteil–Technik*, June, pp. 384–388.

Shannon, R. E. (1975), *Systems Simulation: The Art and Science*, Prentice-Hall, Englewood Cliffs, NJ.

Shapiro, S. S., and Wilk, M. B. (1965), "An Analysis of Variance Test for Normality," *Biometrika*, **52**, 591–611.

Solomon, S. L. (1983), *Simulation of Waiting-Line Systems*, Prentice-Hall, Englewood Cliffs, NJ.

Spaugh, J. M. (1962), "The Use of the Theory of Queues in Optimal Design of Certain Construction Operations," *Technical Report*, National Science Foundation (NSF G 15933), Department of Civil Engineering, University of Illinois, Urbana.

Sprague, C. R. (1972), "Investigation of Hot-Mix Asphaltic Systems by Means of Computer Simulation," Ph.D. dissertation, Texas A&M University, College Station, TX.

Stradal, O., and Cacha, J. (1982), "Time Space Scheduling Method," *J. Construction Division* (ASCE), **108**(C03), 445–457.

Sturges, H. A. (1926), "The Choice of Class Intervals," *J. American Statistical Association*, 65–66.

Swain, J., Venkatraman, S., and Wilson, J. (1988), "Least Squares Estimation of Distribution Functions in Johnson's Translation System," *J. Statist. Comput. Simul.*, **29**, 271–297.

Tavakoli, A. (1983), "Productivity Analysis of Heavy Construction Operations," Ph.D. dissertation, Civil Engineering Department, Georgia Institute of Technology, Atlanta.

Tavakoli, A. (1985), "Productivity Analysis of Construction Operations," *J. Construction Engineering and Management* (ASCE), **111**(1), 31–39.

Tavakoli, A. (1986), "Transient Response Analysis of a Dual Cycle System," *J. Construction Engineering and Management* (ASCE), **112**(3), pp. 403–410.

Teicholz, P. (1963), "A Simulation Approach to the Selection of Construction Equipment," *Technical Report No. 26*, Construction Institute, Stanford University, Stanford, CA, June.

Toomy, R. S. (1984), "The Development of an Earth Distribution Analysis Program for Microcomputers," M.S. Special Topic, School of Civil Engineering, Georgia Institute of Technology, Atlanta.

Touran, A., and Asai, T. (1987), "Simulation of Tunneling Operations," *J. Construction Engineering and Management* (ASCE), **113**(4), 554–568.

Turban, E. (1968), "The Line of Balance—A Management by Exception Tool," *J. Industrial Engineering*, **19**(9), 440–448.

Van Slyke, R. M. (1963), "Monte Carlo Methods and the PERT Problem," *Operations Research*, **1**(3), 839–860.

Venkatraman, S., and Wilson, J. R. (1988), "Modeling Univariate Populations with Johnson's Translation Systems—Description of the FITTR1 Software," *Research Memorandum No. 87-21*, Purdue University, W. Lafayette, IN.

Ward, C. (1985), "Earthwork and Resource Estimation on Large Expedient Projects," *Proceedings, Specialty Conference on Earthmoving and Heavy Equipment* (held at Tempe, AZ), ASCE, New York.

Welch, P. D. (1983), "The Statistical Analysis of Simulation Results," in Lavenberg, S. S. (ed.), *Computer Performance Handbook*, Academic Press.

Willenbrock, J. H. (1972), "Estimating Costs of Earthwork via Simulation," *J. Construction Division* (ASCE), **98**(C01), 49–60.

Wilson, J. R. (1984), "Statistical Aspects of Simulation," *Operations Research*, Elsevier Science Publishers, B.V. (North-Holland), Amsterdam.

Wilson, J. R., and Vekatraman, S. (1987), "Modeling Univariate Populations with the Johnson's Translations System," *Research Memorandum No. 87-21*, Purdue University, W. Lafayette, IN.

Woods, D. G., and Harris, F. C. (1980), "Truck Allocation Model for Concrete Distribution," *J. Construction Division* (ASCE), **106**(C02), 131–139.

Woolery, J., and Crandall, K. (1983), "Stochastic Network Model for Planning Scheduling," *J. Construction Engineering and Management* (ASCE), **109**(3) 342–354.

INDEX